FOOD AND AGRICULTURE ORGANIZATION OF THE UNITED NATIONS

Rome, 2007

Gridded livestock of the world

2007

William Wint and Timothy Robinson

Authors' details
William Wint
Environmental Research Group, Oxford.
Department of Zoology
South Parks Road
Oxford OX1 3PS, UK
william.wint@zoo.ox.ac.uk
Timothy Robinson
FAO Animal Production and Health Division
Rome, Italy
tim.robinson@fao.org.

Recommended citation
FAO. 2007. *Gridded livestock of the world 2007*, by G.R.W. Wint and T.P. Robinson.
Rome, pp 131.

Reprinted 2009

ISBN 978-92-5-105791-9

Contents

Abbreviations and acronyms

AVHRR	Advanced very high resolution radiometer
BTB	Bovine tuberculosis
CIESIN	Centre for International Earth Science Information Network
DEM	Digital elevation model
EROS	Earth Resources Observation Systems
FADS	Farm Animal Demographics Simulator
FAO	Food and agriculture organization of the United Nations
FMD	Foot-and-mouth disease
GAUL	Global administrative unit layers
GIS	Geographic information system
GLC	Global land cover
GLiPHA	Global Livestock Production and Health Atlas
GLIS	Global Livestock Information System
GLW	Gridded livestock of the world
IIASA	International Institute for Applied Systems Analysis
ILRI	International Livestock Research Institute
IUCN	International Union for the Conservation of Nature
LDPS-2	Livestock Development Planning System, Version 2
LGA	Livestock only - arid/semi-arid tropics and subtropics
LGH	Livestock only - humid/subhumid tropics and subtropics
LGP	Length of growing period
LGT	Livestock only - temperate and tropical highlands
MIA	Mixed irrigated - arid/semi-arid tropics and subtropics
MIH	Mixed irrigated - humid/subhumid tropics and subtropics
MIT	Mixed irrigated - temperate and tropical highlands
MODIS	Moderate resolution imaging spectroradiometer
MRA	Mixed rainfed - arid/semi-arid tropics and subtropics
MRH	Mixed rainfed - humid/subhumid tropics and subtropics
MRT	Mixed rainfed - temperate and tropical highlands
NDVI	Normalized Difference Vegetation Index
NOAA	National Oceanographic and Atmospheric Administration
OIE	World Organisation for Animal Health

PATTEC	Pan African Tsetse and Trypanosomiasis Eradication Campaign
SALB	Second administrative level boundaries
TALA	Trypanosomiasis and Land-use in Africa
TLU	Tropical livestock unit
USGS	United States Geological Survey

Foreword

The livestock sector is changing rapidly in response to globalization and the ever-growing demand for animal food products in developing countries, some of which are emerging as powerful new players on the global scene. The expanding trade in livestock and livestock products is constantly under threat from disease outbreaks, thereby calling for better management of transboundary diseases. There are social and environmental consequences of the growth and transformation of this sector: small-scale producers are marginalized and environmental degradation occurs, from both industrial and extensive forms of livestock production; intensification of livestock systems and growing market demands also create a threat to the diversity of animal genetic resources.

Given this dynamic setting, there is a clear need for well-informed livestock sector planning, policy development and analysis, but these are frequently hampered by the paucity of reliable and accessible information on the distribution, abundance and uses of livestock. The FAO Animal Production and Health Division has a global mandate to foster informed decision-making on the challenges facing the livestock sector, particularly those of developing and emerging economies. As a contribution to redressing this shortfall, and in collaboration with the Environmental Research Group Oxford (ERGO), FAO has developed the "Gridded livestock of the world" database: the first standardized global, subnational resolution maps of the major agricultural livestock species. These livestock data are now freely available for download via the FAO Web pages.

The spatial nature of these livestock data allows a wide array of applications. Livestock distribution data provide the units to which parameters may be applied for estimating production; they make it possible to evaluate the impact, both of and on livestock, by applying a variety of rates; and they provide the denominator in prevalence and incidence estimates for epidemiological applications, and identify host distributions for disease transmission models.

Gridded livestock of the world describes how these data have been collected and modelled to produce a digital, geo-referenced global dataset. It also provides varied and extensive examples of some of the applications to which the data have been put. This publication is intended as a point of reference to the data and as a vehicle to stimulate further applications and feedback from those most concerned with the development of the livestock sector – be they policy-makers, researchers, producers or facilitators.

Samuel Jutzi
Director
FAO Animal Production and Health Division

Acknowledgements

A project such as this is obviously the work of more than two people. First and foremost, the authors wish to thank the multitude of people across the world who collected livestock statistics and made them available. None of this would have been possible without such a network of data providers. Joachim Otte has fully supported the project over the last four years, and we are especially grateful to him and David Bourn for their detailed editorial contributions to the manuscript. Gianluca Franceschini manages the GLW database and has formatted the maps included in this publication and extracted the livestock statistics presented in the tables and the annex. Where figures and tables have been derived from other publications the source is clearly indicated; where no source is given the original source is this publication.

A number of people has been involved in this project over the years. At the core of the team, Pius Chilonda, Gianluca Franceschini, Claudia Pitiglio, Federica Chiozza and Valentina Ercoli were involved in the day-to-day data collection and processing; Prof David Rogers and Simon Hay of Oxford University were responsible for the processing and provision of satellite data used to disaggregate the livestock data; Carl Morteo and Adhemar Fontes worked closely with us in developing and implementing the Oracle database; and Pierre Gerber and Tom Wassenaar contributed to the livestock suitability mapping. The artwork in this publication was directed by Claudia Ciarlantini, with contributions from Nicoletta Forlano and James Morgan. Monica Umena was responsible for desktop publishing and Brenda Thomas Bergerre edited the publication.

Not surprisingly, a project such as this has a long history. We are grateful to Jan Slingenbergh and Henning Steinfeld for their initial support and development of livestock geography projects within the Food and Agriculture Organization of the United Nations (FAO).

Many of those mentioned above also contributed to developing the applications presented at the end of this volume. The inputs provided by Russ Kruska, Philip Thornton, Alex Shaw, Marius Gilbert, Guy Hendrickx, Keith Sumption, Freddy Nachtergaele and Ergin Ataman, as well as many other colleagues at FAO and at the TALA Research group at Oxford University who supported this work with advice and ideas, are also gratefully acknowledged.

Summary

One of the major limitations in livestock sector planning, policy development and analysis is the paucity of reliable and accessible information on the distribution, abundance and use of livestock. With the objective of redressing this shortfall, the Animal Production and Health Division of FAO has developed a global livestock information system (GLIS) in which geo-referenced data on livestock numbers and production are collated and standardized, and made available to the general public through the FAO website. Where gaps exist in the available data, or the level of spatial detail is insufficient, livestock numbers are predicted from empirical relationships between livestock densities and environmental, demographic and climatic variables in similar agro-ecological zones.

The spatial nature of these livestock data facilitates analyses that include: estimating livestock production; mapping disease risk and estimating the impact of disease on livestock production; estimating environmental risks associated with livestock due, for example, to land degradation or nutrient loading; and exploring the complex interrelationships between people, livestock and the environment in which they cohabit. It is through quantitative analyses such as these that the impact of technical interventions can be estimated and assessed. Also, by incorporating these data into appropriate models and decision-making tools, it is possible to evaluate the impact of livestock-sector development policies, so that informed recommendations for policy adjustments can be made.

The components of the information system thus created include: a global network of providers of data on livestock and subnational boundaries; an Oracle database in which these data are stored, managed and processed; and a system for predicting livestock distributions based on environmental and other data, resulting in the Gridded Livestock of the World (GLW) initiative: modelled distributions of the major livestock species (cattle, buffalo, sheep, goats, pigs and poultry) have now been produced, at a spatial resolution of three minutes of arc (approximately 5 km). These data are freely available through the GLW website[1], through an interactive web application known as the Global Livestock Production and Health Atlas (GLiPHA)[2], and through the FAO GeoNetwork data repository[3].

As well as detailing various components of the GLIS, this publication explains how livestock distributions were determined, and presents a series of regional and global maps showing where the major ruminant and monogastric species are concentrated.

Spatial livestock data can be used in a multitude of ways. Various examples are given of how these and other datasets can be combined and utilized in a number of applications, including estimates of livestock biomass, carrying capacity, population projections, production and off-take, production-consumption balances, environmental impact and disease risk in the rapidly expanding field of livestock geography.

[1] http://www.fao.org/ag/againfo/resources/en/glw/default.html
[2] http://www.fao.org/ag/aga/glipha/index.jsp
[3] http://www.fao.org/geonetwork/srv/en/main.home

1 Introduction

Livestock make an important contribution to the livelihoods of farming communities and the agricultural economies of most countries. They provide food, fuel and transport, contribute to food security, enhance crop production, generate cash incomes for rural and urban populations, constitute the source of a variety of value-added goods with multiplier effects, and generate a demand for services. Livestock rearing can also diversify production and sources of income, provide year-round employment, spread risk and act as a capital reserve for many agricultural households (FAO, 1996).

On the downside, excessive concentrations of livestock and poorly managed production can have a variety of detrimental impacts on the environment, including: overgrazing, land degradation, nutrient accumulations, water pollution, and greenhouse gas emissions (Bourn et al., 2005). Livestock may have a direct impact on human populations, as they constitute a source of zoonotic diseases.

WHY MAP LIVESTOCK?

Given the economic importance of livestock production, it is essential to have some means of reviewing the relative abundance, and distribution, of livestock resources for the purposes of quantitative analysis, strategic planning and decision support. Maps are a clear and concise way of visualizing large geographical datasets, which would otherwise be difficult to comprehend. They are also an efficient way of storing distribution data and making them easily available for further analysis. Better understanding of the geography of livestock has a variety of potential applications, including:

- determining overall levels of livestock production, and associated feed resource and land requirements;
- quantification and distribution of environmental impacts of livestock production;

- assessing risk from disease, drought, conflict, etc.;
- identifying areas of potential conflict between livestock and crop producers;
- comparing alternative land-use options: arable, mixed, pastoral, ranching, conservation, forestry and tourism, for example;
- assessing the likely impact of technical or policy interventions;
- improving the targeting of livestock-related development initiatives; and
- identifying and quantifying strategic domains (so-called segments) for provision of livestock services, development and disbursement of veterinary pharmaceuticals, etc.

In the wake of the foot-and-mouth disease (FMD) epidemic in the United Kingdom and associated outbreaks in continental Europe in 2001, and the recent emergence of Highly Pathogenic Avian Influenza (HPAI, or bird 'flu) in Southeast Asia, attention has focused on livestock distribution mapping, estimating the numbers of animals at risk of infection, and modelling disease dynamics. A prerequisite for disease-risk mapping is a sound knowledge of the distribution of susceptible species and disease vectors.

LIVESTOCK DIVERSITY

Livestock comprise a broad range of species and breeds of domesticated birds and mammals. Bovines (cattle, buffaloes and yaks) are generally the most highly regarded livestock species because of their size and the quantity, diversity and value of products deriving from them. Bovines are also used for traction and represent major cultural and financial assets in many cultures.

Small ruminants (sheep and goats) may be less highly regarded because of their smaller size and lower value. They are, nevertheless, more numerous and widespread; they breed faster and are more affordable, and are possibly of greater general importance to the poor than are bovines.

Monogastric species (poultry and pigs) are less directly dependent on local land resources for their feed than most other livestock species, and are the mainstay of industrial production systems.

Although resources have not been available to include them within these datasets, the less widespread (camels and yaks) and less numerous (horses, donkeys, mules and asses) species should not be overlooked, because they play a significant role in local rural economies.

The composition of regional and subregional livestock species is likely to change over time in response to the ongoing 'livestock revolution' (Delgado *et al.*, 1999) – the gradual move away from more extensive, land-based, ruminant husbandry to more intensive, short-cycle, monogastric modes of production that are less dependent on local land resources. In some rapidly-growing economies of Asia and South America, these transitions are happening surprisingly quickly.

WHICH FEATURES TO MAP?

In addition to basic population statistics on the numbers of animals within specific administrative areas, a variety of other livestock-related data may be mapped, including:

- numbers and densities;
- species ratios;
- production levels (e.g. of meat, milk, eggs, hides);
- age and sex composition (herd structure parameters);
- constraints to production and causes of mortality;
- livestock diseases;
- productivity parameters and intensification levels;
- levels of trade and prices;
- management and husbandry practices, and ownership; and
- breed distribution and genetic diversity.

The mapping units used, however, must be carefully chosen so as to avoid confusion. For instance, displaying numbers per administrative unit gives a radically different impression to numbers per square kilometre or numbers per person. Expressing animal populations in terms of their weight (biomass) rather than numbers gives a very different perspective again, but allows several species to be combined into a single measurement, such as the tropical livestock unit (TLU), thereby providing some indication of the total quantity of livestock in a specific area.

In general, the availability of these types of information is heavily scale-dependent, and varies widely across the world. Numbers, biomass, production and trade figures are available globally, but usually only at the country level. Herd composition, productivity and socio-economic data tend only to be available for small areas of developing countries, often corresponding to in-depth project area surveys, but may be archived at census-unit level for more developed nations.

Livestock population levels vary in both time and space. Numbers tend to increase with the size of human populations and in concert with cropping levels (Bourn and Wint, 1994), although drought, disease and conflict may severely deplete local livestock populations in the short term. Seasonal movements of stock are also a characteristic feature of drylands and mountainous areas. Livestock productivity and levels of production and consumption also vary, and climate change may be already influencing overall patterns of crop and livestock production. With such inherent variability, it is important to recognize that the maps here presented are composite snapshots derived from the most comprehensive information currently available. These maps may therefore be used as a baseline for future estimations of population change or of the impact of development or other interventions.

2 Disaggregating population data

Livestock data are available in a range of different formats and numerical units: they may be provided as population numbers or densities per square kilometre and are usually presented as summaries, either for the sample unit (e.g. grid estimates for air surveys) or by administrative region (e.g. census units). These different approaches may give rise to rather different-looking maps, as shown in Figures 2.1 and 2.2.

Each approach has a number of advantages and disadvantages: a grid map provides a reasonable representation of a distribution, and can be amalgamated into any number of larger mapping units for comparison with other datasets. There is,

however, the temptation to assign an inappropriate degree of reliability to the figures for an individual grid cell, even if the counts are accurate and precise (which is by no means certain), because populations are rarely static. Administrative (or other) unit maps, on the other hand, are rather inflexible, and manipulation into different mapping units may be difficult. Further, administrative units are forever changing – merging, splitting and shifting boundaries – thereby seriously complicating comparisons between one census and another.

In addition, available data are rarely complete or at a sufficiently high resolution to satisfy the demand from analysts, researchers, policy-mak-

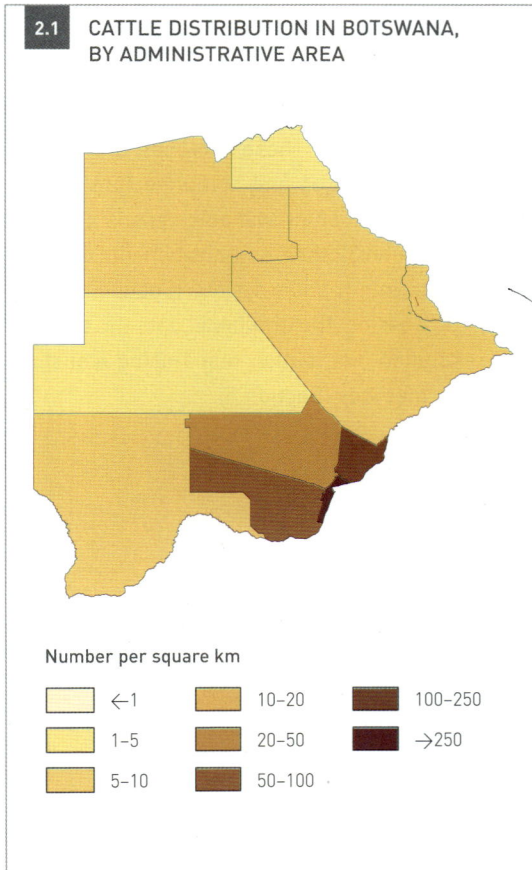

2.1 CATTLE DISTRIBUTION IN BOTSWANA, BY ADMINISTRATIVE AREA

Number per square km

←1	10–20	100–250
1–5	20–50	→250
5–10	50–100	

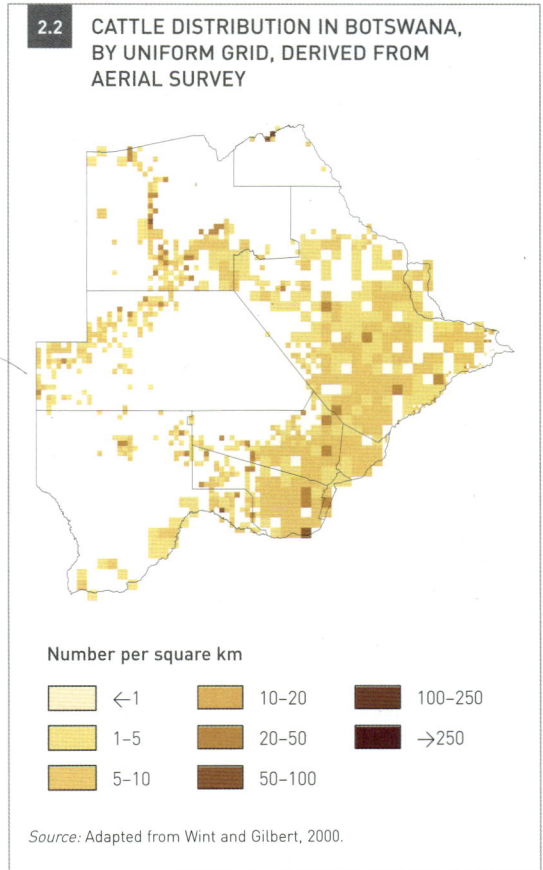

2.2 CATTLE DISTRIBUTION IN BOTSWANA, BY UNIFORM GRID, DERIVED FROM AERIAL SURVEY

Number per square km

←1	10–20	100–250
1–5	20–50	→250
5–10	50–100	

Source: Adapted from Wint and Gilbert, 2000.

ers, etc., for increasingly detailed animal distribution maps. As a result, some form of extrapolation or interpolation is usually needed to provide maps with a complete coverage and standardized format at a useful resolution.

DATA PREDICTION AND EXTRAPOLATION

A number of techniques can be used to enhance available agricultural data.

Interpolation, typified by various Krigging techniques (such as those in the Golden Software's Surfer package[4], in the ESRI ArcGIS Spatial Analyst[5] and in Insightful's S-Plus for the ESRI ArcView Geographic Information System (GIS)[6], may be an appropriate tool for 'improving' point data. However, if meaningful outputs are to be obtained, considerable care is needed when defining various operational parameters (such as search radius and symmetry, degree of smoothing and method selected). Logistic regression or discriminant analysis methods may also be used to 'fill in gaps', but are largely restricted to the use of binary presence/absence or ranked training data that are not usually suitable for estimating population.

Various weighting techniques have also been used to assign national population figures within countries. The least contentious is to 'remove' animals from areas where they can be assumed not to exist (e.g. glaciers, deserts, vertical slopes, tropical rainforest, water bodies and protected areas) and add them to the remaining 'habitable' areas. This 'suitability mapping' approach is discussed in more detail in Section 4.

More ambitious (and thus less assured) methods have utilized the link between domestic livestock and human densities in partitioning national figures for populations (Wint, 1996a), production (Wint, 1996b) and commodities within agro-ecological zones, in accordance with human population levels. This technique can produce serious anomalies, which may be resolved to some extent by refining

the ecological zonations used (White, 1998).

Extrapolation, or distribution modelling, based on established statistical relationship(s) between livestock numbers and a variable, or variables, for which data are available for all the areas of interest, is another possible means for filling data gaps – providing the extrapolation is not taken beyond the value limits of the training data. These, or closely allied, techniques have been used to predict a wide range both of animal distributions, including birds (McPherson *et al.*, 2006) and mammals (Skidmore, 2002) and of arthropod vectors of disease (Rogers *et al.*, 1996; Hay *et al.*, 1996; 2000; 2002; 2006).

FAO has devoted considerable effort to developing this suite of techniques for application at the continental level (e.g. Wint and Rogers, 1998; Wint *et al.*, 1999), which have been extended and enhanced to generate the livestock distribution maps presented in this document. This is the first time such maps have been produced globally and for widespread dissemination in the public domain: it is necessary, therefore, to describe the methods used in some detail. These methods are set out in the following pages and comprise three major stages: the collection of available census and survey data (Section 3); their organization into a standardized data information system (Section 4); and, finally, processing the available data to produce high-resolution distribution maps using statistical modelling methods (Section 5).

---

[4] http://www.goldensoftware.com/products/surfer/surfer.shtml
[5] http://www.esri.com/software/arcgis/extensions/spatialanalyst/index.html
[6] http://www.insightful.com/products/arcview

3 Subnational livestock statistics

The first stage in the mapping process is to collect available subnational livestock statistics, usually for each country. These may be collected and presented in a number of different ways, which can affect the subsequent processing required.

AGRICULTURAL CENSUS METHODS
Livestock data collection methods and frequencies differ according to both their type and economic importance. More detailed and precise information is required for some species than for others, especially where animals' movements need to be traced for compliance with trade regulations or for disease surveillance.

Livestock statistics are usually collected as part of more general censuses of agriculture undertaken periodically by national governments. Agricultural censuses are organized in various ways in different countries, depending upon the resources available, the importance of agriculture, and institutional traditions. Many countries have insufficient resources to mount a series of detailed surveys for different parts of the agricultural sector and thus restrict their efforts to obtaining data from a single agricultural census, every five to ten years. Such censuses may involve complete or sample coverage, with the agricultural holding as the standard unit of enumeration. It should be noted, however, that many agricultural censuses do not include animals located in communal grazing areas or fallow land under shifting cultivation (FAO, 1995a), both of which may be important categories in many (particularly developing) countries.

The first World Census of Agriculture took place in 1930 under the auspices of the former International Institute for Agriculture in Rome. A follow-up census planned for 1940 was prevented by World War II, after which FAO took on responsibility for promoting and coordinating a regular world census of agriculture that has taken place every ten years since 1950, most recently in 2000 (FAO, 1995b). While FAO has actively promoted the standardization of agricultural census procedures and livestock data collection[7], considerable variation remains in the detail and reliability of national statistics. Livestock statistics are not restricted to numbers: censuses often also assess herd structure, production parameters, and information on marketing and trade.

The collection of livestock statistics is a national government responsibility that is usually associated with obtaining more general agricultural statistics, and should be standardized as far as possible in terms of species, breed and product categories, and units of measurement. The importance attached to the collection of agricultural statistics and thus the resources allocated to this activity, however, vary from country to country.

Livestock censuses are usually conducted by ground-based surveys and questionnaires, often of sample households, and frequently in conjunction with censuses of arable agriculture or, occasionally, agro-economic surveys. Census techniques vary from country to country, depending on circumstances. In countries such as the United Kingdom and the United States, for instance, agricultural census information is obtained directly from farmers, who are required by law to provide information requested in periodic, postal questionnaires. This is effective as long as the great majority of farmers receive and understand the questionnaires, and are willing to provide the information requested. However, this methodology relies on comprehensive registration of owners, if not the animals themselves. And in many less developed countries, where formal registration of farms and farmers is often limited to the commercial sector, this method of postal census

[7] http://www.fao.org/es/ess/rmlive.asp

is clearly inappropriate as it would not only exclude the majority of small-scale, rural farmers but would also require the existence of a functional postal system and universal literacy.

UNDER-REPRESENTATION

The basic unit of enumeration for most, if not all, agricultural censuses is the 'agricultural holding'. Areas of communal grazing, fallow land and shifting cultivation are usually excluded. Unless, in census design, special provision is made to offset this inherent bias in favour of permanent, fixed landholdings, most agricultural statistics will inevitably under-represent the livestock holdings of nomadic and transhumant pastoralists with 'no fixed abode'. This under-representation of pastoral livestock is a considerable problem in under-populated, higher rainfall areas such as the sub-humid zone of West Africa, but is likely to be particularly significant in arid and semi-arid regions of Africa, Asia and South America, large areas of which are, at the best of times, relatively remote and inaccessible; Norton-Griffiths, 1978, for example, makes reference to systematic under-estimation of nomadic livestock.

It is also important to recognize that many developing countries do not have adequate means of collecting, analysing and reporting agricultural (or, indeed, human) population statistics. Available information about cropped areas and livestock resources is, therefore, often incomplete and of uncertain reliability. On its FAOSTAT web site[8], FAO acknowledges that *"... many developing countries still do not have an adequate system of statistics pertaining to the agricultural sector. Some of the available agricultural data are incomplete [and] even when data are available, their reliability may be questionable."* It is for this reason that alternative means of assessing land cover and livestock resources need to be used for remote and inaccessible regions of many developing countries, especially in Africa.

Low-level aerial surveys, originally developed to count wildlife (Norton-Griffiths, 1978), have been widely used to assess livestock populations in many countries across Africa (Clarke, 1986; Government of Kenya, 1996). These have been further developed to incorporate ground survey methods in order that a range of livestock species can be assessed: from larger ruminant and monogastric species to domestic pigeons and beehives. Such direct counting methods may produce markedly different results to those provided by census methods that rely on stakeholder responses. The 1990 National Livestock Census of Nigeria, which pioneered air-ground census techniques, indicated that there were substantially more livestock than estimated by the Federal Office of Statistics: twice as many cattle; one and a half times as many sheep and goats; and four times as many pigs (Bourn *et al.*, 1994).

DATA SUPPRESSION

A frequent problem for the agricultural statistician is that many countries, particularly those in the industrialized world that conduct holding-level censuses, are constrained by data protection and confidentiality legislation to suppress data that could allow an individual holding to be identified. As a result, many data records for the less numerous species, or for those that are restricted to few large holdings within a mapping unit (e.g. industrialized pig or poultry production units), may be withheld from census statistics released in the public domain. Ironically this means that public domain agricultural statistics from the United Kingdom and the United States, for example, may contain more gaps than data from developing countries.

[8] http://faostat.fao.org/

4 FAO global livestock information system

Any global archive of subnational livestock data is required to satisfy a number of criteria. Data must be checked and validated to minimize errors and omissions and, where necessary, be converted into standard parameters and units so that information from various sources can be compared. To maintain its usefulness the archive must be regularly and easily updated; sources and procedures must, therefore, be properly documented, catalogued and automated.

The structure of the FAO livestock information data archive and its processing protocols are described below. Subsequent subsections describe the procedures used to apply supplementary information to enhance the raw data and treat missing data, and explain the exclusion, or masking out, of areas known to be incapable of supporting livestock.

DATA ARCHIVE STRUCTURE AND PROCESSING

For many years, FAO has collated and distributed national-level data on livestock and related commodities through the well-known FAOSTAT database. More recently, however, efforts have been made to systematize the collection, management, processing and distribution of subnational livestock data. This was originally carried out at the administration level 1 (usually the province) through the GLiPHA project, and more recently at the highest available spatial resolution in support of the GLW initiative. Figure 4.1 provides a schematic summary of the information system.

Underpinning the information system is a growing network of providers of subnational livestock data. The sources of data are very diverse and include statistical yearbooks, development project

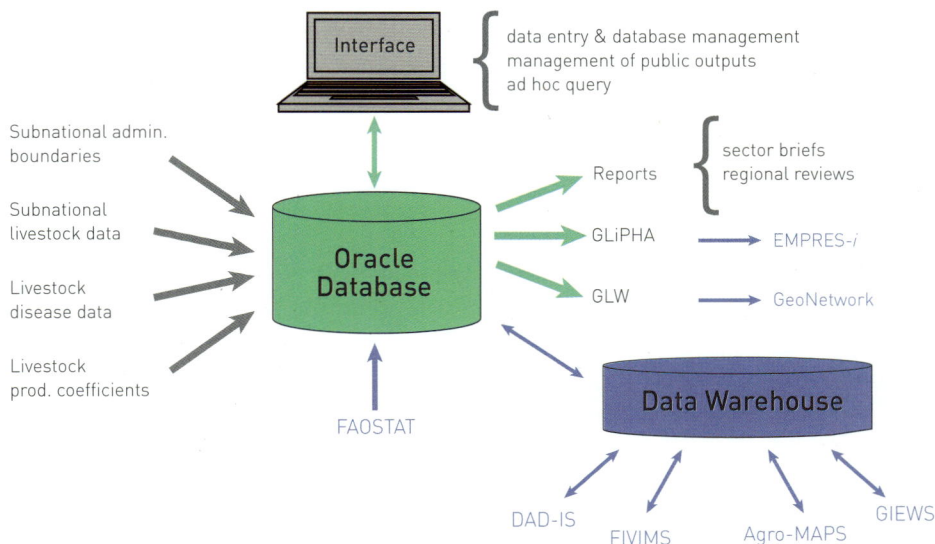

4.1 SCHEMA OF THE FAO GLOBAL LIVESTOCK INFORMATION SYSTEM

documents, contacts within national departments and an increasing number of sources of livestock data that are available over the Internet. Indeed, even over the four-year development of these distribution data, the rise in official web pages has been remarkable. A database of national partners responsible for livestock statistics, together with website hyperlinks, is maintained for the purpose of providing feedback and value-added data products. Hand-in-hand with the livestock data is geo-referenced information on subnational boundaries. This is sometimes provided with the livestock data but, more usually, different departments are responsible for producing and maintaining these geographic data. This means that the livestock statistics need to be matched with the available administrative data, based on administrative unit names or codes. There are various initiatives to standardize national and subnational boundary data and codes, which are used wherever feasible. The United Nations Geographic Information Working Group of the United Nations Cartographic Service maintains a well-documented dataset of international boundaries and areas under dispute[9], which is used for national boundaries. Two global initiatives exist for standardized subnational boundaries: the World Health Organization's Second Administrative Level Boundaries (SALB) project[10] and the FAO Global Administrative Unit Layers (GAUL) project[11]. These two systems are related but differ in important ways. The SALB datasets, the first initiative to standardize subnational boundaries globally, are only provided to the second administrative level (the national boundary being level zero), and are standardized to the year 2000 and endorsed by the national cartographic units. This slows down the process significantly and tends to restrict coverage. The GAUL system was designed to 'fast track' these procedures and therefore boundaries are not

formally endorsed; thus it is not in the public domain but currently restricted to United Nations use. GAUL uses the most recently available boundary data and makes use of whatever resolution is available. To allow rapid updating of boundaries, it has also adopted a more versatile coding system. The FAO livestock information system originally adopted the SALB coding system and used SALB data where available, upgrading it with more recent and more detailed data as needed and available. As new national livestock statistics become available and are entered into the system, however, the GAUL standards will be adopted. Livestock disease data are restricted to the national-level World Organisation for Animal Health (OIE) Handistatus II[12] and supplemented by national reports that provide some subnational resolution information. The OIE is now finalizing the World Animal Health Information System, which will replace Handistatus, and is collating subnational livestock disease data. This new resource will be used once it becomes operational. Livestock performance indicator values from published and grey literature are currently maintained in separate databases.

Once acquired, the raw livestock and boundary data are digitized and managed via a web-based interface to an Oracle database. A number of data verification procedures are embedded, including a direct link to the FAOSTAT database[13] from which country totals are compared against FAO 'official' statistics.

There are various outputs from the primary database. These include ad hoc queries and standardized tables of statistics and maps that are published in FAO's national livestock sector briefs, which provide livestock sector profiles for specific countries and regional livestock sector reviews. A major component of the global livestock information system is GLiPHA[14], an interactive web application that draws livestock and socio-

[9] http://boundaries.ungiwg.org
[10] http://www.who.int/whosis/database/gis/salb/salb_home.htm
[11] http://www.fao.org/geonetwork/srv/en/metadata.show?id=12691

[12] http://www.oie.int/hs2
[13] http://faostat.fao.org/
[14] http://www.fao.org/ag/aga/glipha/index.jsp

economic data from the Oracle database, usually at the first administrative level (province). Data are compiled into national and regional 'projects' and can be viewed and downloaded as tables, graphs and maps, with raster backdrops of layers such as elevation and vector overlays of roads, population centres and other relevant features. GLiPHA also feeds directly into the EMPRES-*i* database[15], where detailed disease outbreak data can be overlain on the standard livestock and other GLiPHA layers.

A further output from the database is to the FAO 'data warehouse', a recent concept within the organization designed to bring together many of the disparate databases and information systems available in-house. The underlying principle is that a standardized spatial coding system is adopted, by which links are established to data and data products that are likely to be of particular relevance to other departments within FAO. These data items are assigned thematic codes and regularly updated by drawing on the most recent statistics from the participating information systems. The data warehouse concept is at an early stage of development and is being piloted by the GLIS project and the Global Information and Early Warning System[16], with interest from other information systems such as the Food Insecurity and Vulnerability Information and Mapping System[17], DAD-IS[18] (an information system on animal genetic resources) and Agro-MAPS[19] (an information system on crop-based agriculture).

The main topic of this publication, however, and indeed the reason for developing the GLIS, is the new GLW. For this output, the most recent livestock statistics in the Oracle database are extracted at the highest available spatial resolution to feed into the GLW analysis chain. The following sections provide a detailed description of the processing involved in producing the GLW datasets.

[15] http://www.fao.org/ag/aga/agah/empres/tadinfo/e_tadh.htm
[16] http://www.fao.org/es/giews/english/index.htm
[17] http://www.fivims.net
[18] http://www.fao.org/dad-is
[19] http://www.fao.org/landandwater/agll/agromaps/interactive/page.jspx

SUPPLEMENTARY AND MISSING DATA

Census and survey records are often incomplete, with gaps that need to be filled to provide complete maps. Various methods have been devised to generate credible estimates of missing data.

There are, for instance, many areas where the number of animals present is known, or can be safely assumed, to be zero – either from country-level statistical records, such as FAOSTAT, or because of a cultural prohibition such as the ban on pigs in most Islamic countries. Known zeros can also derive from land suitability masking, in which areas unsuitable for specific types of livestock are defined according to various climatic, demographic and topographic criteria: for example, cattle do not usually live in deserts or the middle of rainforests. The definition of suitable land is discussed below.

In some instances, particularly for less common species, only country-level population figures are available – often from FAOSTAT – because census summary data, or yearbooks, do not include subnational figures. These can be treated by assigning animal numbers to administrative areas according to the land area of the units, or by weighting the assignment of numbers by some other relevant parameter, such as human population, for which administrative-level data are known. Use of human population distribution to apportion livestock populations is often most appropriate for poultry and pigs, which, in developing countries, are closely associated with human populations. In such manipulations, administrative-level data, rather than pixel values, are used to assign polygon densities. Human population must then be excluded from the suite of predictors used in any subsequent distribution modelling (Section 5).

Complete, subnational population datasets for all livestock species are not available for all countries. Some have administrative-level data available for only part of the country because of incomplete enumeration or data suppression to ensure confidentiality.

These incomplete datasets can be often rectified by using data available for a higher administrative level. For example, if data for administrative level 2 are available for part of a country and data for level 1 are known, subtraction of known level-2 totals from level-1 totals will give the number of animals in the region for which level-2 data are not available. A single density can then be calculated for the level-2 administrative areas, or numbers can be assigned in relation to an associated parameter, as previously mentioned.

It should be emphasized, however, that the adjustments described in the preceding paragraphs should not be applied to very large polygons unless the area of land deemed suitable for a given species in that polygon is comparatively small.

MASKING LAND SUITABLE FOR LIVESTOCK

Deserts, lakes and high mountains are unsuitable for either arable or livestock production. Cultivation and animal husbandry are also not usually allowed in national parks or game reserves. Such factors must obviously be taken into account in producing livestock distribution maps, in which densities indicate the number of animals per square kilometre of land suitable for livestock production rather than simply the total land area.

Input criteria

Areas known to be unsuitable for livestock must be defined and delineated using standard criteria that can be applied globally, so that animal densities in those areas can be set to zero.

Land suitability criteria for two broad categories – (i) rainfed crop cultivation and ruminant livestock production (cattle, buffaloes, sheep and goats); and (ii) monogastric livestock production (pigs and chickens) – have been defined in terms of a number of globally available spatial variables, as described and explained below.

Protected areas

Depending on their classification and the level of enforcement, protected areas generally exclude livestock. The International Union for the Conservation of Nature (IUCN) protected area

categories I-IV were considered unsuitable for livestock. Categories V and above, which include, for example, forest reserves that are frequently used by livestock, particularly in the developing world, were not excluded. The IUCN database is becoming increasingly comprehensive[20] but has been supplemented by the Managed Areas Database for North America[21] and national data for South Africa, Botswana and Kenya.

Infrastructure and demography

Cities were also defined as unsuitable, using demographic layers derived from the LandScan coverages[22] rather than the Gridded Population of the World[23], which had not been finalized by the time the GLW coverages were first generated. Both population density and night-time lights were included, albeit with very high thresholds, because it became apparent that each had been used to define urban areas, but in different ways in different locations. These high thresholds delineated areas that corresponded well, though not precisely, with the developed and partly developed LandScan land-cover categories[24], which were also incorporated.

Closed canopy forest

A variety of digital layers of forest cover are available in the public domain, the most recent being the University of Maryland's 500 m resolution percentage tree cover[25], derived from Moderate Resolution Imaging Spectroradiometer (MODIS) satellite imagery, and the Global Land Cover (GLC) 2000[26] forest layers under development at the European Commission's Joint Research Centre at Ispra, Italy. When compared with the earlier 1 km resolution layers derived from Advanced Very High Resolution Radiometer (AVHRR) imagery[27], it was evident that closed forest, as defined in the GLC

[20] http://www.iucn.org/themes/wcpa
[21] http://www.geog.ucsb.edu/~gavin/mad/mad.html
[22] http://www.ornl.gov/sci/gist/projects/LandScan
[23] http://sedac.ciesn.columbia.edu/gpw
[24] http://www.ornl.gov/sci/gist/projects/LandScan
[25] http://www.glcf.umiacs.umd.edu/data/treecover
[26] http://www-gvm.jrc.it/glc2000/
[27] http://www.glcf.umiacs.umd.edu/data/treecover

TABLE 4.1 DATASETS AND THRESHOLDS USED TO DETERMINE LAND UNSUITABLE FOR LIVESTOCK

Criteria[1]	Map Layer	
	Rainfed agriculture and ruminant livestock production[2]	Monogastric livestock production[3]
Protected areas (1/0)	1	1
Population density (Landscan) (km^{-2})	$\rightarrow$ 1 500	$\rightarrow$ 1 500
Lights (Landscan) (%)	$\rightarrow$ 90	$\rightarrow$ 90
Slope (Landscan) (%)	$\rightarrow$ 40	-
Elevation (m)	$\rightarrow$ 4 750	$\rightarrow$ 4 750
Pasture suitability (IIASA) (% area)	0	-
NDVI max	$\leftarrow$ 0.07	-
Tree cover - South America (Maryland GLCF) (%)	$\rightarrow$ 75	-
Tree cover - rest of world (MODIS) (%)	$\rightarrow$ 95	-
Land cover (Landscan) – water (1/0)	1	1
Land cover (Landscan) – developed (1/0)	1	1
Land cover (Landscan) – partly developed (1/0)	1	1
Land cover (Landscan) – wetlands (1/0)	1	1
Land cover (Landscan) – wooded wetlands (1/0)	1	1
Land cover (Landscan) – tundra (1/0)	1	1
Land cover (Landscan) – snow and ice (1/0)	1	1

[1] The datasets used are described and referenced in the text (Section 4.3).
[2] Cattle, buffalo, sheep and goats.
[3] Pigs, chickens and other poultry.

2000 coverage, extended over a much larger area than other coverages, particularly in Southeast Asia. It was also apparent that MODIS estimates were more homogenous and considerably higher than corresponding AVHRR values, at least for the Amazon Basin. As a very conservative definition of forest cover was required, MODIS coverage was used in preference to GLC 2000 in all regions except South America, for which the Maryland AVHRR values were used.

Climate

It was initially assumed that land suitable for livestock could be identified from estimated air temperatures derived from the AVHRR satellite imagery of the National Oceanic and Atmospheric Administration (NOAA) (United States). However, regions with very high minimum or mean tem-peratures – for example, much of the Sahel – are known to support livestock for at least part of the year. Maximum temperatures were also seen as ineffective discriminators, as they excluded large parts of China and Patagonia, for example, which are known to support significant numbers of rumi-nants. Temperature was thus excluded from the suitability criteria used.

Topography

Threshold values for elevation (derived from the global GTOPO30 1 km resolution Digital Elevation Model [DEM], produced by the United States Geological Survey [USGS], Earth Resources Observation Systems [EROS] data centre[28]) and slope (derived from layers in the LandScan archive[29]), were set

[28] http://edc.usgs.gov/products/elevation/gtopo30/gtopo30.html
[29] http://www.ornl.gov/sci/gist/projects/LandScan

4.2 ESTIMATED LAND UNSUITABLE FOR RUMINANT LIVESTOCK PRODUCTION IN AFRICA

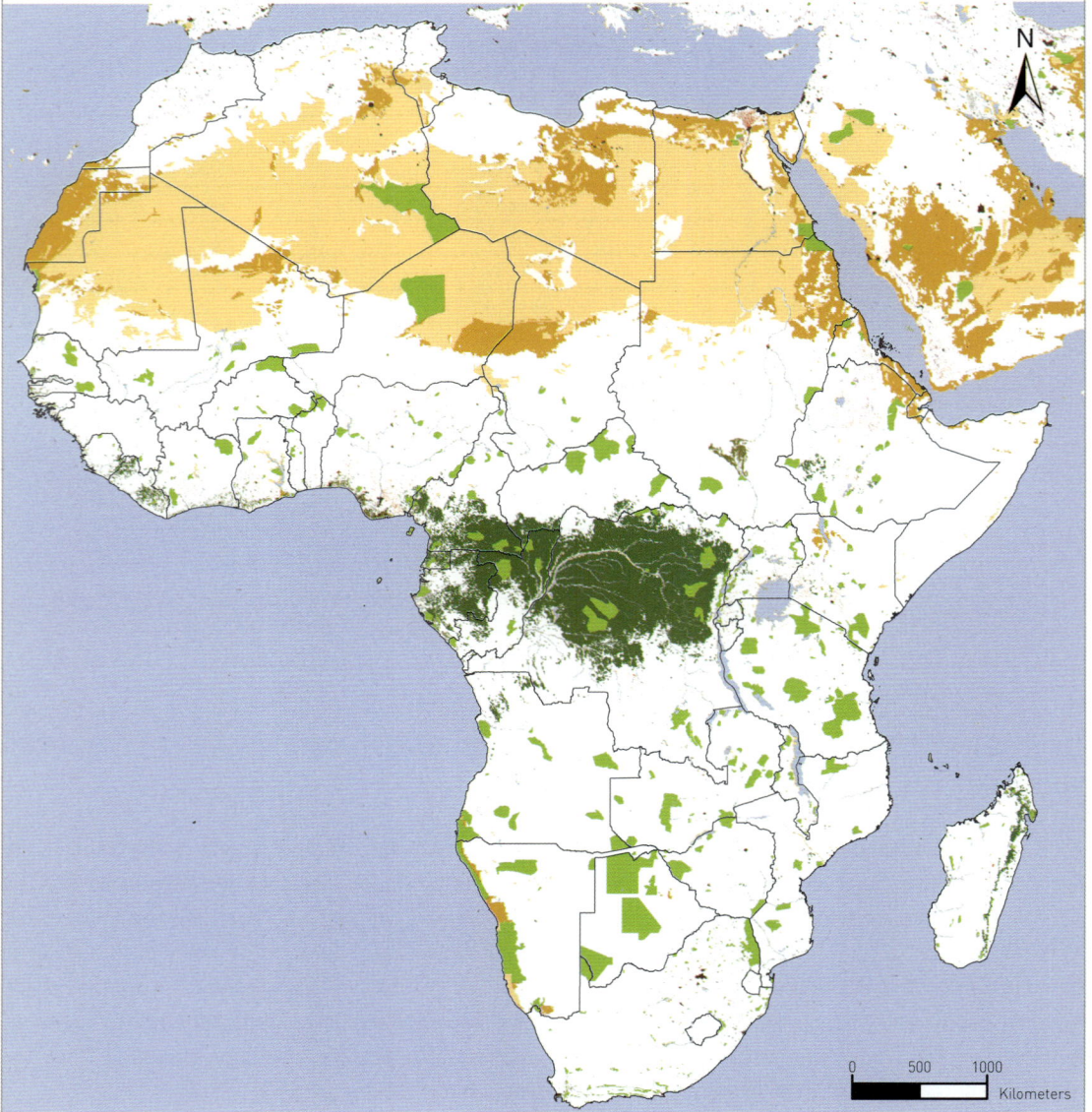

Unsuitable criteria for ruminants

International boundary

Water

Land cover: developed, wetlands, snow and ice

Protected areas

Night-time lights →90%

Elevation →4750 m

Tree cover →95%

Pasture suitability 0

NDVI maximum →0.07

Human population density →1 500 square km

Slope →40 degree

to exclude the highest peaks in the Himalayas and Andes, and pixels with extremely high slope values.

Vegetation

Satellite-derived vegetation greenness, the Normalized Difference Vegetation Index (NDVI) (Green and Hay, 2002; Hay, 2000; Hay *et al.*, 2006), working maps of pasture suitability provided by the International Institute for Applied Systems Analysis (IIASA) and estimated land cover categories, derived from the LandScan land cover dataset[30], were all considered as potential determinants of land suitability. Apart from the urban categories (see above), only the most inhospitable land cover categories were excluded – water, wetland, cold tundra and snow, or ice – as even the lowest vegetation category (barren) included places in the Near East and the Sahel known to support ruminants. For the same reason, only pixels defined as unsuitable for rainfed pasture (with a score of zero) were deemed unsuitable for livestock.

Maximum NDVI was considered a better indicator of vegetation cover than mean values, on the assumption that land with a very low maximum cover would rarely, if ever, be suitable for livestock, whereas areas with a low mean value could be seasonally well-vegetated and therefore support livestock at some times of the year. Thresholds for maximum NDVI, land cover and pasture suitability were based on the arid Near East, where detailed analyses had been conducted previously (Wint, 2003).

Thresholds and results

It was assumed that subsequent regression procedures incorporated in distribution modelling (Section 5) would help to locate marginally unsuitable areas, as well as those where the boundary values varied from region to region. Each threshold, therefore, was conservatively defined to ensure that this process of thresholding excluded only the most unsuitable land. Each parameter was examined in regions with which the analysts were familiar and thresholds subsequently selected, as set out in Table 4.1.

The estimated extent of land unsuitable for rainfed crop and ruminant livestock production in Africa is given in Figure 4.2 as an example, showing the contribution made by the different criteria to the overall suitability mask.

---

[30] http://www.ornl.gov/sci/gist/projects/LandScan

5 Modelling livestock distribution

Once the available agricultural statistics have been collected, standardized, enhanced with supplementary data and adjusted for the extent of land deemed suitable for livestock production, the resulting data archive provides a sound basis for statistical distribution modelling. This process depends on establishing a robust statistical relationship between livestock numbers and one, or more, predictor variable for which data are available for the entire area of interest. These relationships are detailed later in this section.

The modelling process, including inputs and outputs at the various stages, is summarized in Figure 5.1. This process relies on the use of raster images to store both observed (or training) data (i.e. livestock densities) and all the predictor variables.

Statistical relationships are established between observed livestock densities and predictor variables using values extracted for a series of regularly spaced sample points, as illustrated in Figure 5.2. The resulting equations are then applied to all pixels in the predictor images so as to produce a predicted distribution map.

As a result, the technique can be used to predict livestock densities in areas for which no livestock data are available, i.e. filling in gaps. Moreover, because predicted densities are produced at the resolution of the raster imagery, the models generate heterogeneous densities within polygons that have only one single observed value, thus disaggregating the original data. For limited datasets, therefore, the method has the major advantage of

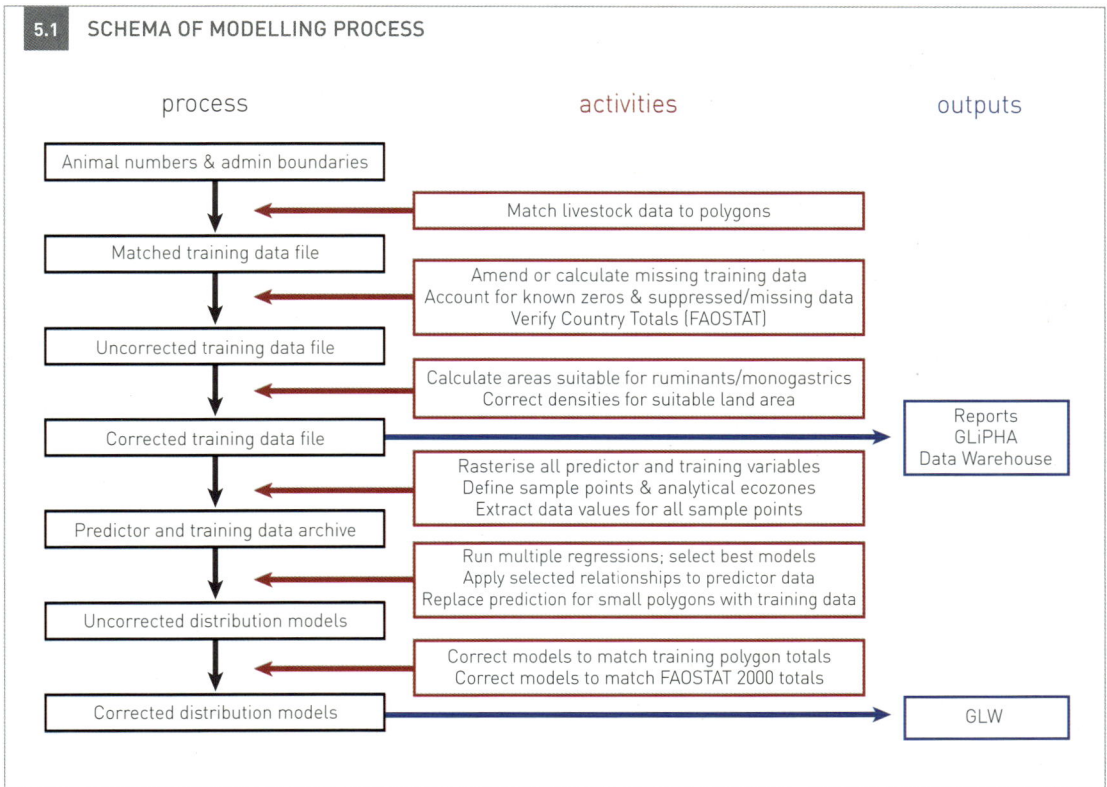

5.1   SCHEMA OF MODELLING PROCESS

process · activities · outputs

- Animal numbers & admin boundaries
  - Match livestock data to polygons
- Matched training data file
  - Amend or calculate missing training data
  - Account for known zeros & suppressed/missing data
  - Verify Country Totals (FAOSTAT)
- Uncorrected training data file
  - Calculate areas suitable for ruminants/monogastrics
  - Correct densities for suitable land area
- Corrected training data file → Reports GLiPHA Data Warehouse
  - Rasterise all predictor and training variables
  - Define sample points & analytical ecozones
  - Extract data values for all sample points
- Predictor and training data archive
  - Run multiple regressions; select best models
  - Apply selected relationships to predictor data
  - Replace prediction for small polygons with training data
- Uncorrected distribution models
  - Correct models to match training polygon totals
  - Correct models to match FAOSTAT 2000 totals
- Corrected distribution models → GLW

5.2   SCHEMA OF SAMPLING PROCESS

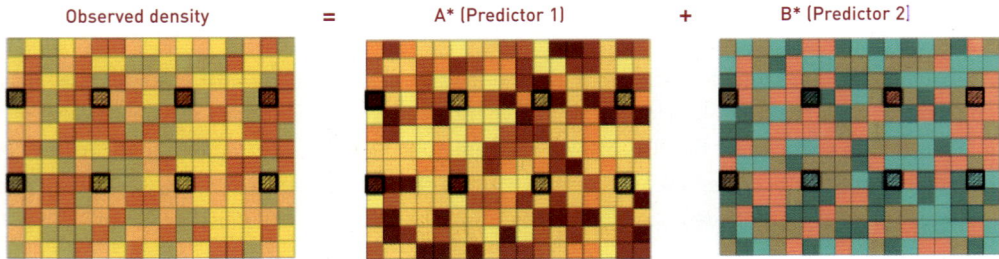

Observed density   =   A* (Predictor 1)   +   B* (Predictor 2)

1) Convert all data maps to images with the same pixel size (resolution);
2) Extract values for observed values of livestock density, and for each predictor variable at fixed sample points (hatched squares);
3) Calculate a regression equation of the form:
   Observed density = Constant + A * (Predictor 1) + B * (Predictor 2) + ...;
4) Providing the equation is statistically significant (i.e. reliable), apply the right hand side of the equation to **all** pixels in the predictor variable images to produce the predicted density;
5) Repeat the process for each of a series of analysis zones (e.g. ecozones).

---

both filling in gaps and refining the level of detail that can be mapped.

As the predictors of animal density are unlikely to be consistent from region to region, the modelling process should be run at several different spatial scales to provide a range of predictive relationships appropriate to specific areas. As well as administrative-level analyses an ecological stratification has been routinely used, on the assumption that the factors determining animal distributions are likely to be similar in areas with comparable ecological characteristics, thereby allowing (i) more robust statistical relationships between training data and predictor variables to be established, and (ii) more realistic predictions of livestock densities in other parts of the same ecological zone for which data are not available.

The ecological zones used to stratify the modelling were defined separately for each continent using non-hierarchical clustering techniques, either within the ADDAPIX programme (Griguolo and Mazzanti, 1996) or ERDAS Imagine software (Leica Geosystems®). The input parameters were drawn from the suite of predictor variables and

included elevation and a series of remotely sensed parameters (the mean [Fourier component 0] and phase [Fourier component 1] of middle infrared, land surface temperature, vegetation index, air temperature and vapour pressure deficit). See below for further details.

A WORKED EXAMPLE - AFRICA

The whole modelling process can be illustrated using cattle in Africa as an example. Observed cattle densities were derived from various national census reports, livestock surveys and data archives for the period 1992-2003. As can be seen in Figure 5.3, most known, or 'observed', cattle densities relate to administrative units, some of which are very large.

A series of stepwise multiple regression analyses was performed to establish the statistical relationships between observed cattle densities and a range of predictor variables drawn from those described below, including: satellite-derived measures of rainfall, temperature, vapour pressure deficit, vegetation cover and elevation (provided by the Trypanosomiasis And Land-use in Africa

5.3 OBSERVED CATTLE DENSITIES IN AFRICA

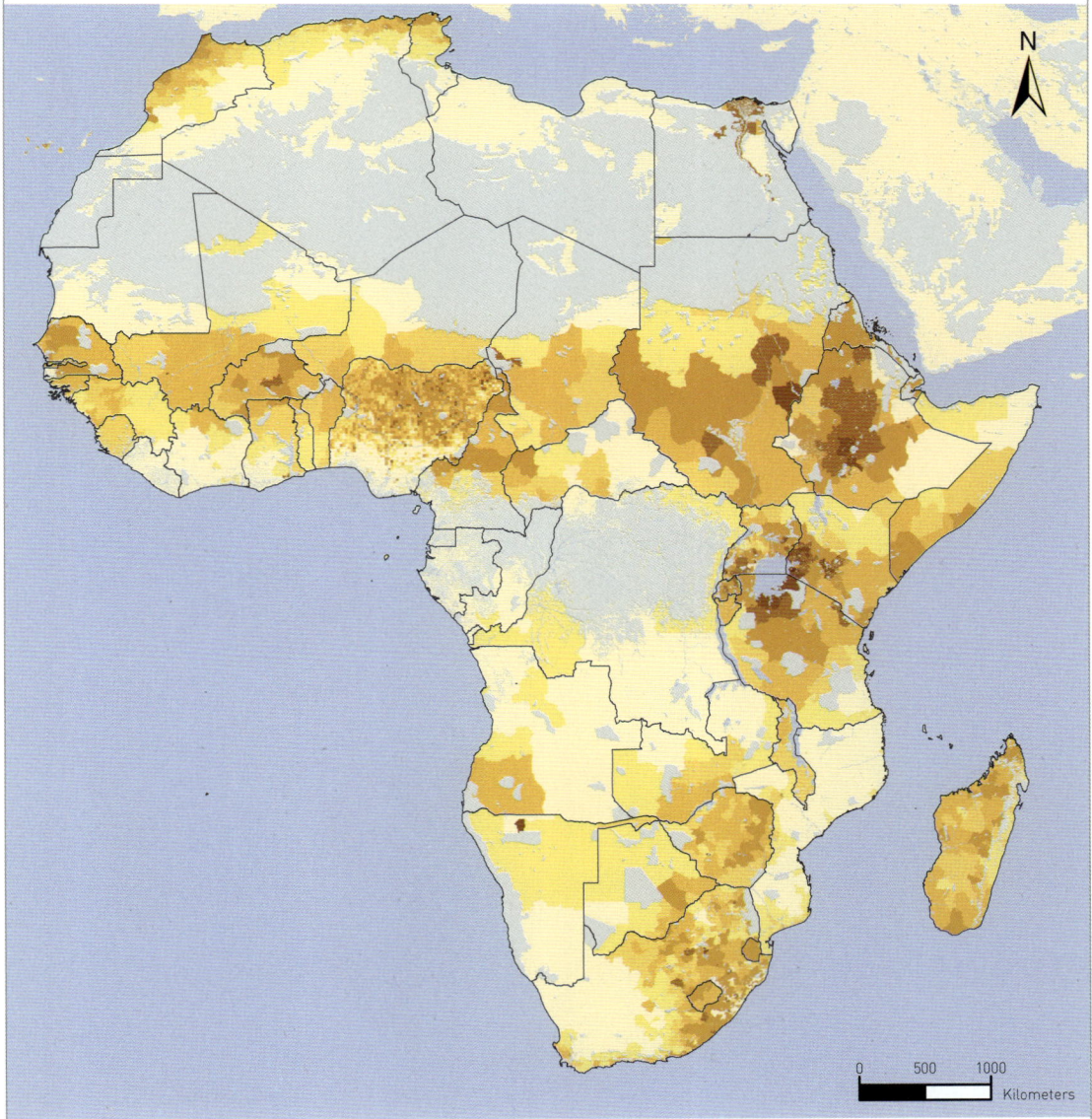

Number per square km

←1	10–20	100–250
1–5	20–50	→250
5–10	50–100	

International boundary
Water
Unsuitable for ruminant

Source: Robinson *et al.*, 2007.

5.4 MODELLED CATTLE DENSITIES IN AFRICA

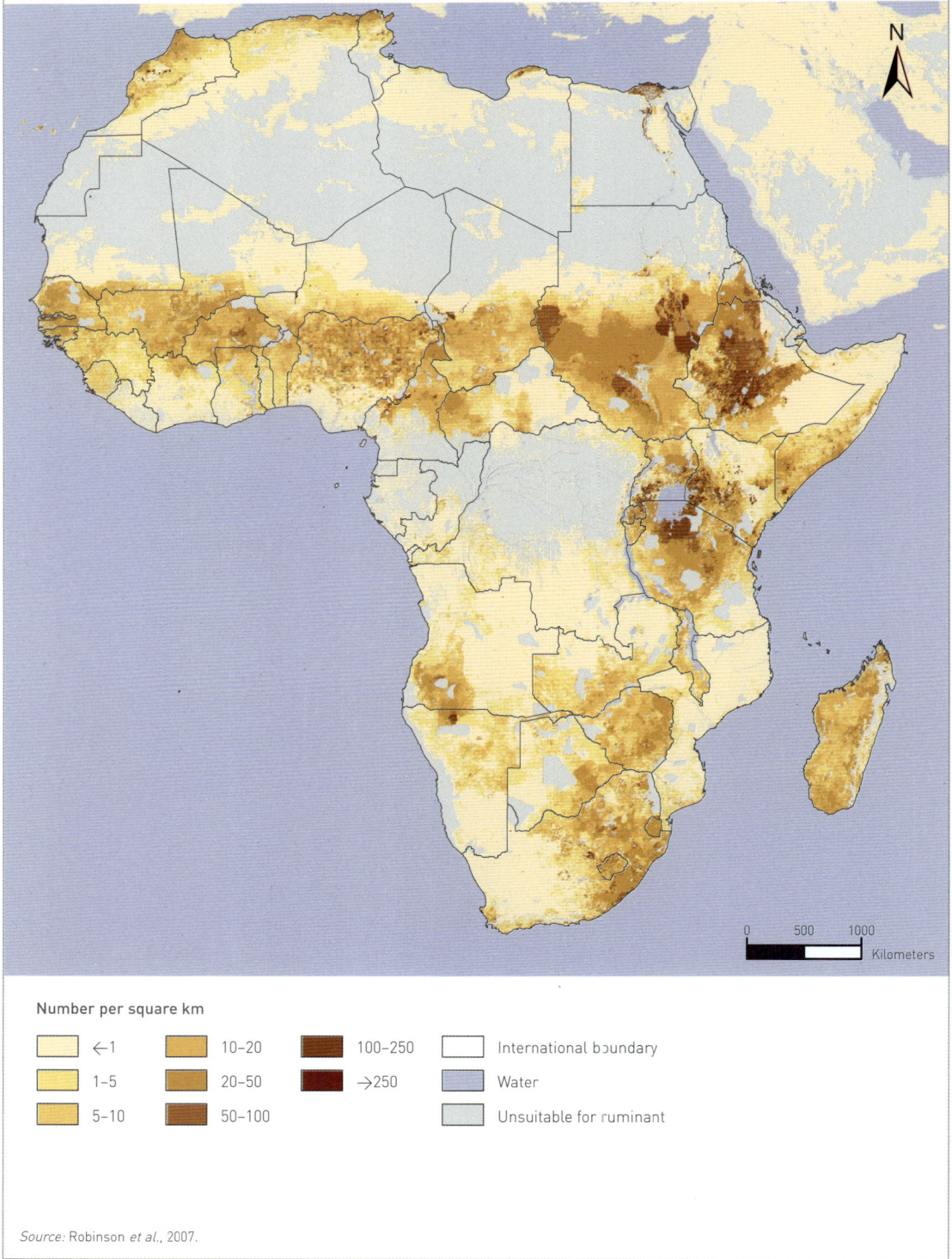

Number per square km

←1	10–20	100–250
1–5	20–50	→250
5–10	50–100	

International boundary
Water
Unsuitable for ruminant

Source: Robinson *et al.*, 2007.

(TALA) Research Group at the University of Oxford); potential evapotranspiration; length of growing period (LGP); human population; and the potential number of tsetse species present.

Values were extracted for approximately 19 000 regularly spaced sample points and a series of regression models derived at different spatial scales: (i) the entire continent; (ii) four continental subregions (east, west, south and north); (iii) 50 ecological zones; and (iv) each ecological zone within each country. In addition, four sets of transformations were assessed – no transformation, logarithmic, exponential and power – to address the possibility that relationships were non-linear. The best relationship was then selected based on R^2 values. Approximately 500 equations were generated and assessed, of which some 60 were selected to contribute to the model. Where statistically valid equations were not found at the highest spatial resolution, equations for the next spatial scale up (region) were assessed, and so on until an acceptable model was identified for each unit of area.

The resulting equations were then applied to the original imagery to generate a map of predicted cattle distribution at a spatial resolution of three minutes of arc (approximately 5 km^2 at the equator). All the predictive equations used were statistically significant at the 1 percent level (p←0.01), or better; but it is axiomatic that the validity of the predicted distribution map depends primarily on the accuracy of the 'observed' training data.

The predicted cattle distribution in Africa, as shown in Figure 5.4, mirrors the observed distribution (Figure 5.3) very well and picks out both major foci (e.g. the Kenya, Ethiopia and Zimbabwe highlands, Tanzania, semi-arid and dry subhumid West Africa), as well as smaller concentrations such as the Gezira irrigation scheme in Sudan, the inland delta of the Niger River in Mali and southeastern Zambia. Relatively high-resolution observed data for Nigeria, derived from aerial survey, were smoothed by the regression analysis. Some of the contrasts between observed and predicted maps are due to minor differences in values falling into different mapping classes. There are also some minor anomalies in northern Chad, where very high predicted densities are obviously false and are caused by extreme predictor values. Human population density is a major determinant of cattle distribution in Africa (Bourn and Wint, 1994) and was the primary predictor in 30 percent of regression equations used.

There is, of course, a danger that these predictive techniques, based on intensive processing, may conceal substantial errors; it is all too easy to be seduced by the fact that a somewhat messy map of fairly reliable data has been converted into an aesthetically more pleasing one, with no holes and apparently believable content. Validation is also problematic because verification is likely to be based on original polygon data rather than by commissioning new survey data, which is time-consuming and expensive. Thus, any variation generated within the polygon (a primary objective for the prediction in the first place) will be seen as a deviation from known data and may, therefore, be considered erroneous. On the other hand, pixel-by-pixel comparisons are equally invidious and error-prone, as the predictions used are statistically based and designed to be interpreted en masse rather than individually. This suggests that a high-resolution prediction can be effectively validated only when re-compiled to administrative-level summaries.

To minimize inconsistencies between original records and summed predicted values, two sets of standardized outputs have been produced in addition to the previously described 'raw' predictions.

STANDARDIZING PREDICTED DISTRIBUTIONS

The numerical outputs of distribution modelling generally had mean values per polygon similar to those of the training data, but rarely matched exactly because regression analysis tends to smooth the peaks and troughs. In addition to the standardization imposed by the suitability masking, the following standardization procedures were adopted.

- model predictions for small polygons – defined as less than 1 000 km^2 – were replaced by suitability-corrected training data;
- model distributions were corrected so that totals calculated for training polygons matched the input training data, referred to as 'totals-corrected' distributions; and
- model values were adjusted so that calculated national totals matched the FAOSTAT country populations for year 2000, the so-called 'year 2000-corrected' distributions.

These corrections involved calculating a ratio between predicted and training data values for each polygon of observed (training) data and then applying the inverse of that ratio to the predicted data densities. The exception was where training data were absent, in which case predicted values were left unchanged.

Of the three routinely produced versions of livestock distribution based on suitability-corrected observed data, suitability- and totals-corrected, and suitability- and year 2000-corrected, the suitability- and totals-corrected version is the preferred output and is the version presented in the next section.

PREDICTOR VARIABLES
A wide range of parameters has been incorporated in the analysis and modelling procedures, including ecoclimatic data, topography, human population data, cartographic data and data on protected areas.

Satellite imagery
The livestock distribution modelling used the following satellite-derived measures of land-surface and atmospheric characteristics:

- NDVI from the AVHRR; a widely accepted measure of vegetation cover (Green and Hay, 2002; Hay, 2000; Hay et al., 2006). Data were provided by the Pathfinder Program, initially supplied by the United States National Aeronautics and Space Administration's Global Inventory Monitoring and Modelling

Systems group;
- a measure of land surface temperature derived by the TALA research group from thermal channels 4 and 5 of the AVHRR using the Price split window technique (Green and Hay, 2002; Hay, 2000; Hay et al., 2006; Hay and Lennon, 1999; Price, 1984);
- a measure of air temperature (Tair), also derived from AVHRR channels (Goetz et al., 2000);
- a measure of middle infrared reflectance, allied to temperature but less susceptible to atmospheric interference, derived from channel 3 of the AVHRR data (Hay, 2000);
- a measure of vapour pressure deficit derived from AVHRR channels 4 and 5 and ancillary processing (Green and Hay, 2002; Hay, 2000; Hay et al., 2006); and
- a surrogate for rainfall – 'cold cloud duration' – derived from Meteosat remotely sensed data, provided by the FAO Artemis data archives (Hay, 2000).

All satellite-derived data were available as a series of decadal (ten-day) composite images, the AVHRR data covering an 18-year period from 1982 to 2000 and the Meteosat data covering a 29-year period from 1961 to 1990. Each series was subjected to temporal Fourier processing (named after the French mathematician, Joseph Fourier), re-sampled to 0.05-degree resolution (approximately 5 km^2 at the equator) and re-projected to the latitude/longitude system (geographic or Plate Carrée projection). The Fourier processing of satellite data, described in detail in Rogers and Williams, 1994; Rogers et al., 1996; Rogers, 1997; and Rogers, 2000, is quite central to the modelling process since it reveals the all-important seasonal characteristics of the environment. Each multi-temporal series is reduced to seven separate data layers: the mean, and the phases and amplitudes of the annual, bi-annual and tri-annual cycles of change. These are supplemented by three additional variables: the

TABLE 5.1 GENERIC LIST OF VARIABLES USED IN LIVESTOCK DISTRIBUTION MODELLING

Generic type	Variables
Locational	Longitude, latitude
Anthropogenic	Distance to roads [1]
	Distance to city lights [1]
Demographic	Human population [2]
Topographic	Elevation [3]
Land cover	Normalized difference vegetation index [4, 5, 6]
Temperature	Land surface temperature [4, 5, 6, 7, 8]
	Air temperature [9]
	Middle-infrared [5]
Water and moisture	Vapour pressure deficit [4, 5, 6]
	Distance to rivers [10]
	Cold cloud duration [5, 11]
	Potential evapotranspiration [11]
General climatic	Modelled length of growing period [12]
Other	Tsetse distributions (for Africa) [13]

[1] Derived from layers in the LandScan archive, produced and distributed by Oak Ridge National Laboratories (ORNL) (http://www.ornl.gov/sci/gist/projects/LandScan).
[2] Taken from CEISIN's Gridded Population of the World (GPW) version 2 dataset (http://sedac.ciesin.columbia.edu/gpw).
[3] Global GTOPO30 1km resolution elevation surface, produced by the Global Land Information System (GLIS) of the United States Geological Survey, Earth Resources Oservation Systems (USGS, EROS) data centre (http://edc.usgs.gov/products/elevation/gtopo30/gtopo30.html).
[4] Green and Hay, 2002.
[5] Hay, 2000.
[6] Hay et al., 2006.
[7] Hay and Lennon, 1999.
[8] Price, 1984.
[9] Goetz et al., 2000.
[10] Derived from the USGS EROS data centre HYDRO 1k data archive (http://edc.usgs.gov/products/elevation/gtopo30/hydro/index.html).
[11] Mean, minimum and maximum decadal estimates of 'cold cloud duration' were derived from METEOSAT remotely-sensed data (1961-90), obtained from FAO's Artemis data archives.
[12] Fischer et al., 2002 (http://www.fao.org/waicent/faoinfo/agricult/agl/agll/gaez/index.htm).
[13] Tsetse distributions used were those developed for the Programme Against African Trypanosomiasis (PAAT) Information system (http://www.fao.org/ag/againfo/programmes/en/paat/infosys.html).

Source: Robinson *et al.,* in press.

maximum, the minimum and the variance of the satellite-derived geophysical variables.

The temporal Fourier processing of multi-temporal data is illustrated in Figures 5.5 (land surface temperature) and 5.6 (NDVI), both taken from FAO, 2006b. In each case, three years of monthly AVHRR data are shown as the black lines (the additional grey line in year 1 is the three-year average). The annual, bi-annual and tri-annual Fourier cycles are shown in red, green and blue, respectively (notice the second, zero-centred scale for these on the upper graph, right-hand axis),

and their re-combined sum is shown as the violet line super-imposed on the raw data. These figures illustrate how the Fourier decomposition manages to capture subtle details of the seasonal cycle in both variables.

The Fourier variables were calculated and turned into GIS image data layers, together with the maximum and minimum values and variances of each original signal. Collectively, these numerical indictors of the level (mean, minimum, maximum), timing (phase), seasonality (amplitude) and variability (variance) of each satellite-derived

5.5 FOURIER-PROCESSED LAND SURFACE TEMPERATURE TIME SERIES FROM A SINGLE POINT IN NORTHERN CÔTE D'IVOIRE

Khorogo, Cote d'Ivoire, West Africa, c 6W, 10N

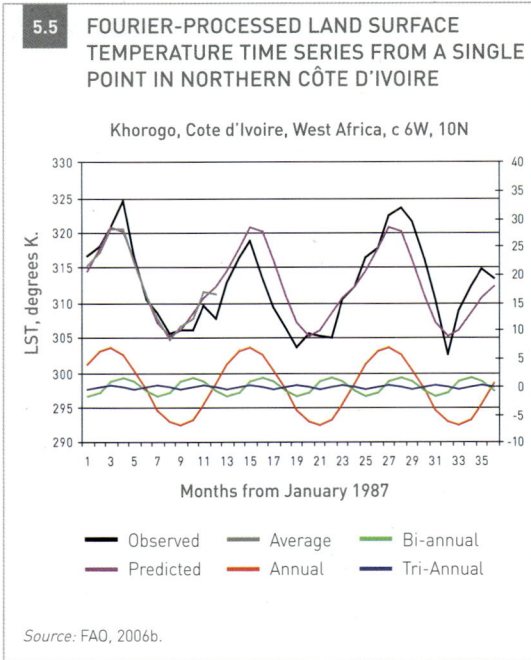

Months from January 1987

— Observed   — Average   — Bi-annual
— Predicted   — Annual   — Tri-Annual

Source: FAO, 2006b.

5.6 FOURIER-PROCESSED NDVI TIME SERIES FROM A SINGLE POINT IN NORTHERN CÔTE D'IVOIRE

Khorogo, Cote d'Ivoire, West Africa, c 6W, 10N

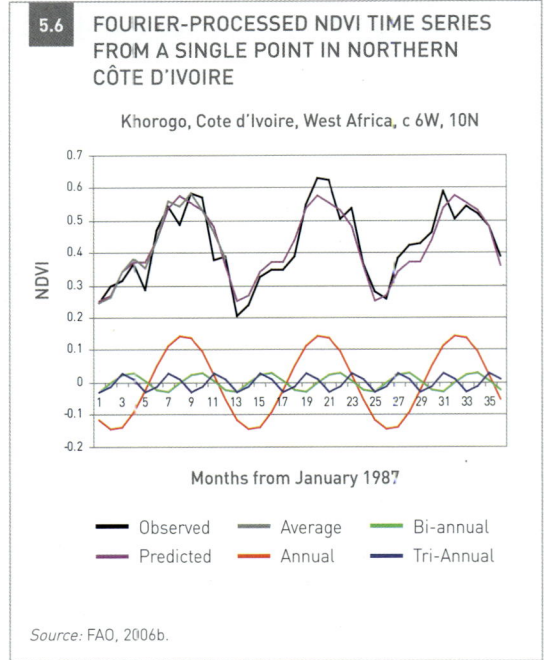

Months from January 1987

— Observed   — Average   — Bi-annual
— Predicted   — Annual   — Tri-Annual

Source: FAO, 2006b.

environmental variable give a unique 'fingerprint' of habitat type; they provide a link between the satellite signal and the biological processes that are, in one way or another, linked to the suitability of the environment to support livestock. A further advantage of the Fourier processing is that it reduces the vast number of individual decadal images to a manageable and relatively independent set of variables, more amenable to statistical analysis and interpretation.

The power of these Fourier-processed data to distinguish habitat types is illustrated in Figure 5.7, taken from Rogers and Robinson, 2004, in which three of the Fourier variables for the NDVI images for Africa are combined as a false colour composite: the average value (or 'zero-order' component) is displayed in red; the phase of the first-order component is displayed in green; and the amplitude of the first-order component is displayed in blue.

Other eco-climatic and land-related data

Elevation data were obtained from the USGS EROS data centre's GTOPO30 1 km resolution DEM

for Africa[31]. A series of land-use variables were extracted from the LandScan data set[32], including slope and vegetation cover. In addition, rivers were taken from the USGS EROS data centre's HYDRO 1k data archive[33]. Larger rivers were identified according to their flow accumulation characteristics, from which a distance-to-rivers image was prepared.

Potential evapotranspiration and annual rainfall data were taken from the FAO/IIASA data archive (Fischer *et al.*, 2002)[34] and re-sampled to a 0.05-degree resolution.

The LGP was modelled separately for each continent, using regression techniques illustrated earlier in this section and the FAO/IIASA archive values as training data.

Human population data

As the GLW project has evolved, so also have the sources of human population data used in the modelling. Early on, for Africa and Asia, human

[31] http://edc.usgs.gov/products/elevation/gtopo30/gtopo30.html
[32] http://www.ornl.gov/sci/gist/projects/LandScan
[33] http://edc.usgs.gov/products/elevation/gtopo30/hydro/index.html
[34] http://www.fao.org/waicent/faoinfo/agricult/agl/agll/gaez/index.htm

5.7 FALSE COLOUR COMPOSITE OF FOURIER-PROCESSED NDVI VARIABLES FOR AFRICA

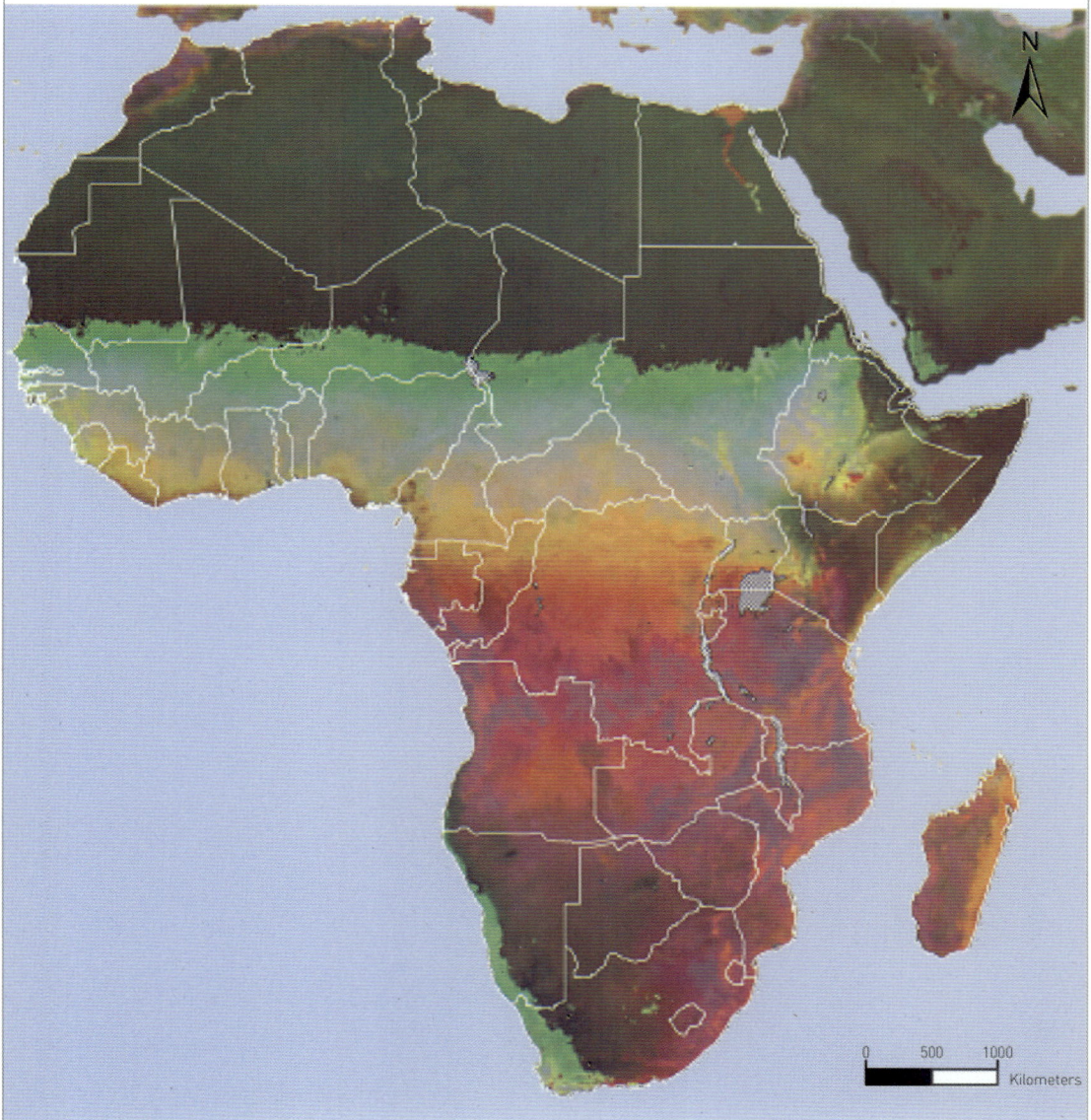

Source: Rogers and Robinson, 2004.

population data were derived from three sources: (i) estimates collated by the FAO Agriculture Land and Water Division at five-minute resolution; (ii) data, again at five-minute resolution, provided by the Centre for International Earth Science Information Network (CIESIN), derived from data collated by the National Centre for Geographic Information and Analysis[35]; and (iii) data for the Horn of Africa came from the Intergovernmental Authority on Drought and Development – now known as the Intergovernmental Authority on Development (Wint et al., 1997). Pixel values from these sources were averaged.

[35] http://www.ncgia.ucsb.edu/pubs/gdp/pop.html

More recently efforts have been made to compile global human population data: first, the LandScan project[36] and, more recently, CIESIN's Gridded Population of the World[37], which is now in its third version and includes the Global Rural-Urban Mapping Project datasets. In the more recent analyses, the project has moved towards these more consistent datasets.

Other data related to human population distributions and proximity to night-time lights and roads were generated from layers available in the Columbia University LandScan archive[38].

[36] http://www.ornl.gov/sci/gist/projects/LandScan
[37] http://sedac.ciesin.columbia.edu/gpw

[38] http://www.ornl.gov/sci/gist/projects/LandScan

Results

The modelled livestock distributions are now available globally, in regional tiles, for a wide range of species, as summarized in Table 6.1.

A representative selection of the numerous livestock distribution maps generated by the GLW project is presented in this section. This is the first example of global subnational transboundary distributions produced using a consistent methodology for all species. Cattle have been quite frequently mapped subnationally, at continental scale, but this is the first time such maps have been produced at this resolution for small ruminant and monogastric species.

DISTRIBUTION OF BOVINE SPECIES

Figure 6.1 shows the global distribution of bovine species. This is dominated by the distribution of cattle over most of the world, but in Asia represents the combined distribution of cattle and buffalos.

Cattle are fairly ubiquitous, except in the very high latitudes where it is too cold for them to survive and in deserts and rainforests where no food is available. India stands out as having by far the largest population, with other global foci in north-

west Europe, the east African highlands and the Sahel, and parts of Brazil and Argentina. Figures 6.2, 6.3 and 6.4 show the modelled cattle distributions in South America, Australasia and Europe, respectively; as seen above, the cattle distributions for Africa are given in Figures 5.3 (observed) and 5.4 (modelled).

Buffaloes are primarily concentrated in India, with significant densities in Southeast Asia and the Philippines, and very limited numbers elsewhere. Figure 6.5 shows the distribution of buffaloes in Southeast Asia.

DISTRIBUTION OF SMALL RUMINANT SPECIES

At the global level, small ruminants are less widespread than bovine species (Figure 6.6). Major sheep populations occur in the Near East (Figure 6.7), throughout Australasia, in the United Kingdom and in southern Brazil. There is also a continuous band of comparatively high density stretching from Spain and northwest Africa (Figure 6.8) through to northwest India. Unlike cattle, sheep tend to have a more restricted distribution within the larger coun-

TABLE 6.1 SUMMARY OF GLW DATA BY SPECIES AND REGION

Species Group	Regional Tile							
	AFR	AMN	AMS	ASE	ASW	AUS	EUR	FSU
Cattle	•	•	•	•	•	•	•	•
Buffaloes				•	•			
Sheep	•	•	•	•	•	•	•	•
Goats	•	•	•	•	•	•	•	•
Pigs	•	•	•	•	•	•	•	•
Chickens	•	•	•	•	•	•	•	•
Poultry				•				

Notes:
AFR = Africa; AMN = North America and the Caribbean; AMS = Latin America; ASE = East and Southeast Asia;
ASW = West Asia & the Middle East; AUS = Australasia; EUR = Europe; and FSU = the former Soviet Union.

tries or regions: the African Sahel, South Africa, southern India, north-central China and Mongolia, for example.

Goats are more localized than sheep and often very restricted in their distribution, for example, to southern Texas in the United States and to the northeast of Brazil. Figure 6.9 illustrates how their distribution in the former Soviet Union is concentrated around eastern Turkmenistan and Uzbekistan, western Tajikistan and Kyrgyzstan, and southern Kazakhstan. Goats are widespread in Africa and more common than sheep in the Sahel and East Africa (Figure 6.10). Major foci also occur in India, Pakistan, Indonesia, north-central China and the Near East.

DISTRIBUTION OF PIG SPECIES
Figure 6.11 shows the distribution of pigs at the global level. Distribution of this particular livestock species is that most influenced by religious and cultural factors. There are few or none of these animals in predominantly Islamic countries such as Pakistan (Figure 6.12). Large populations occur in eastern China and parts of Southeast Asia, such as Viet Nam (Figure 6.12), Western Europe, central and eastern areas of the United States (Figure 6.13), Central America (Figure 6.13) and southern Brazil.

DISTRIBUTION OF POULTRY SPECIES
Figure 6.14 shows the global distribution of poultry. Massive numbers of birds occur in China and Southeast Asia (Figure 6.15), but Europe also has large poultry populations. Whereas in areas where poultry are abundant (particularly in rural settings) human population is by far the strongest predictor variable, the global distribution is not wholly related to human population distribution. India, for example, though densely populated, does not have a correspondingly ubiquitous poultry population. Some countries with large absolute populations appear not to have a widespread distribution of poultry, probably because very large numbers of birds are concentrated in small areas or in individual production units that are not easily visible at continental resolution.

6.1 GLOBAL BOVINE DISTRIBUTION (Modelled)

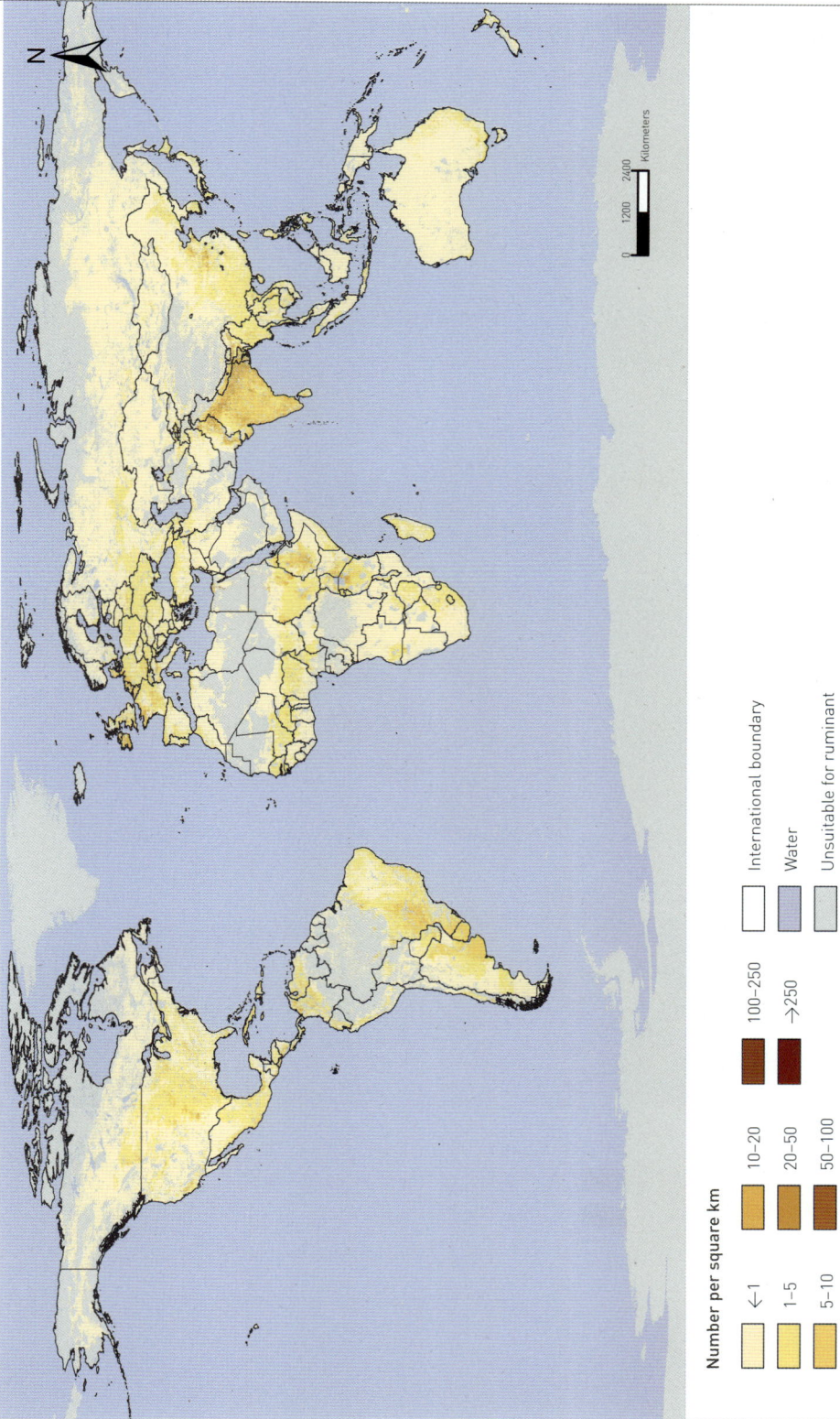

Number per square km

<1	10–20	100–250
1–5	20–50	→250
5–10	50–100	

International boundary

Water

Unsuitable for ruminant

0 1200 2400 Kilometers

6.2 DISTRIBUTION OF CATTLE IN SOUTH AMERICA (Modelled)

Number per square km

←1	10–20	100–250
1–5	20–50	→250
5–10	50–100	

International boundary

Water

Unsuitable for ruminant

6.3 DISTRIBUTION OF CATTLE IN AUSTRALASIA (Modelled)

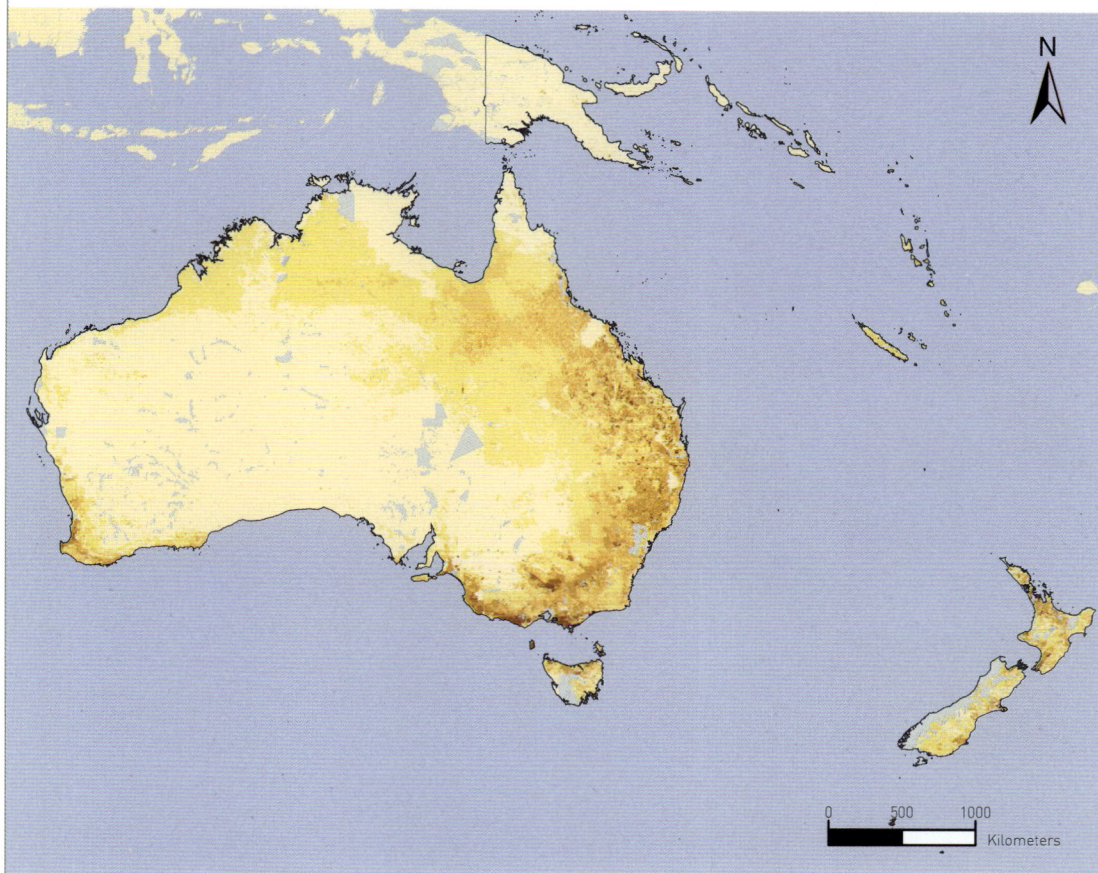

N

0   500   1000
Kilometers

Number per square km

←1	10–20	100–250
1–5	20–50	→250
5–10	50–100	

International boundary

Water

Unsuitable for ruminant

6.4 DISTRIBUTION OF CATTLE IN EUROPE (Modelled)

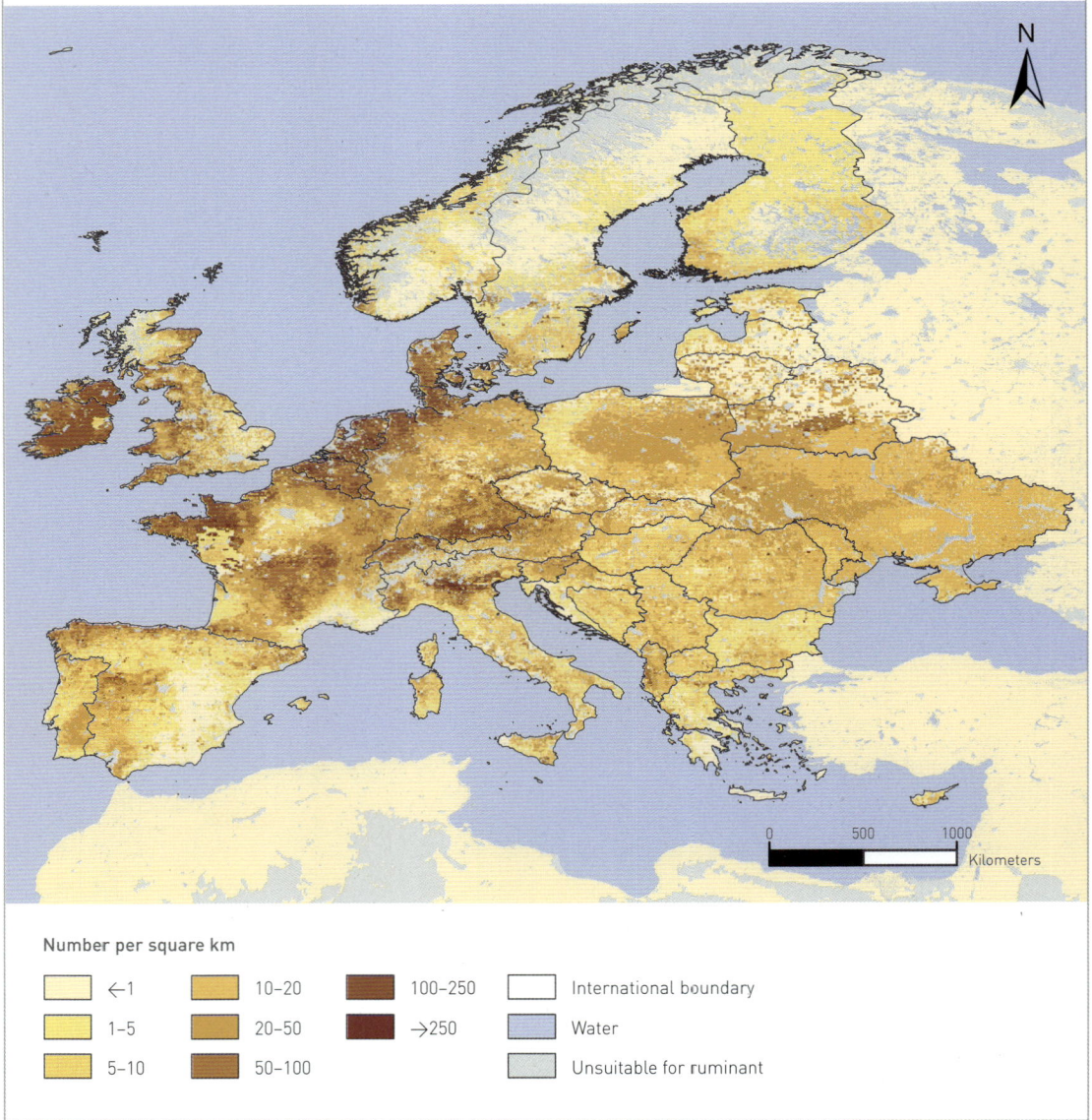

Number per square km

←1	10–20	100–250	International boundary
1–5	20–50	→250	Water
5–10	50–100		Unsuitable for ruminant

6.5 DISTRIBUTION OF BUFFALOES IN EAST AND SOUTHEAST ASIA (Modelled)

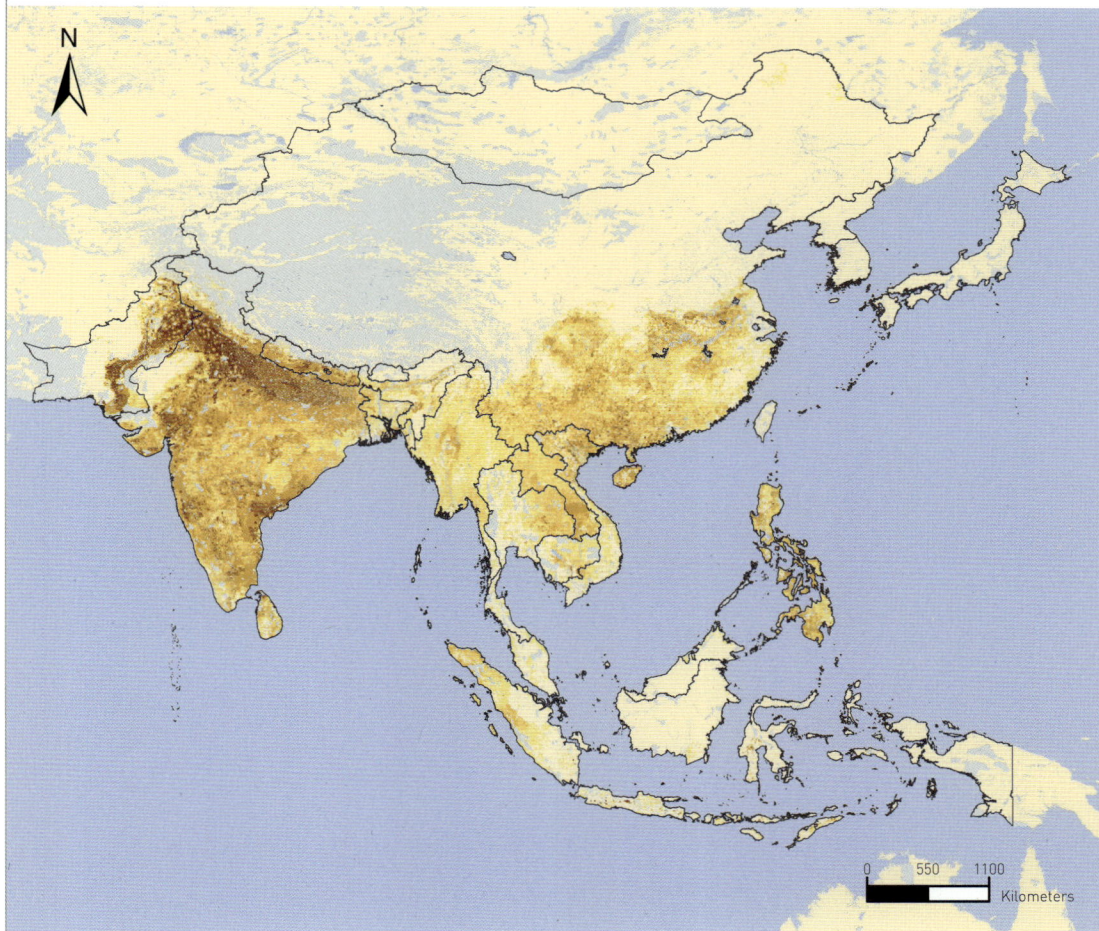

Number per square km

←1	10–20	100–250
1–5	20–50	→250
5–10	50–100	

International boundary

Water

Unsuitable for ruminant

6.6 GLOBAL SMALL RUMINANT DISTRIBUTION (Modelled)

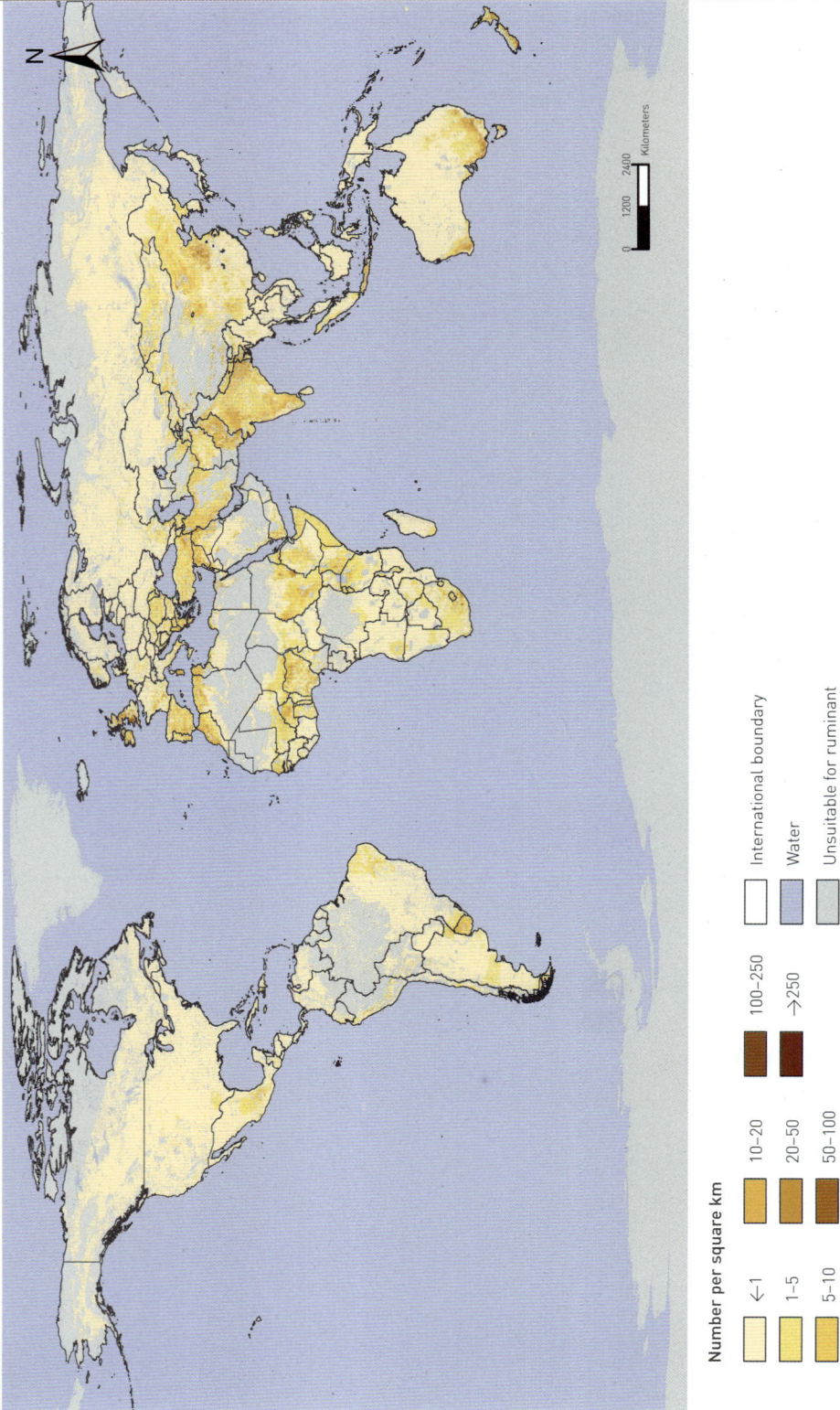

Number per square km

<1	10–20
1–5	20–50
5–10	50–100
	100–250
	→250

International boundary

Water

Unsuitable for ruminant

Kilometers
0   1200   2400

6.7 DISTRIBUTION OF SHEEP IN THE NEAR EAST (Modelled)

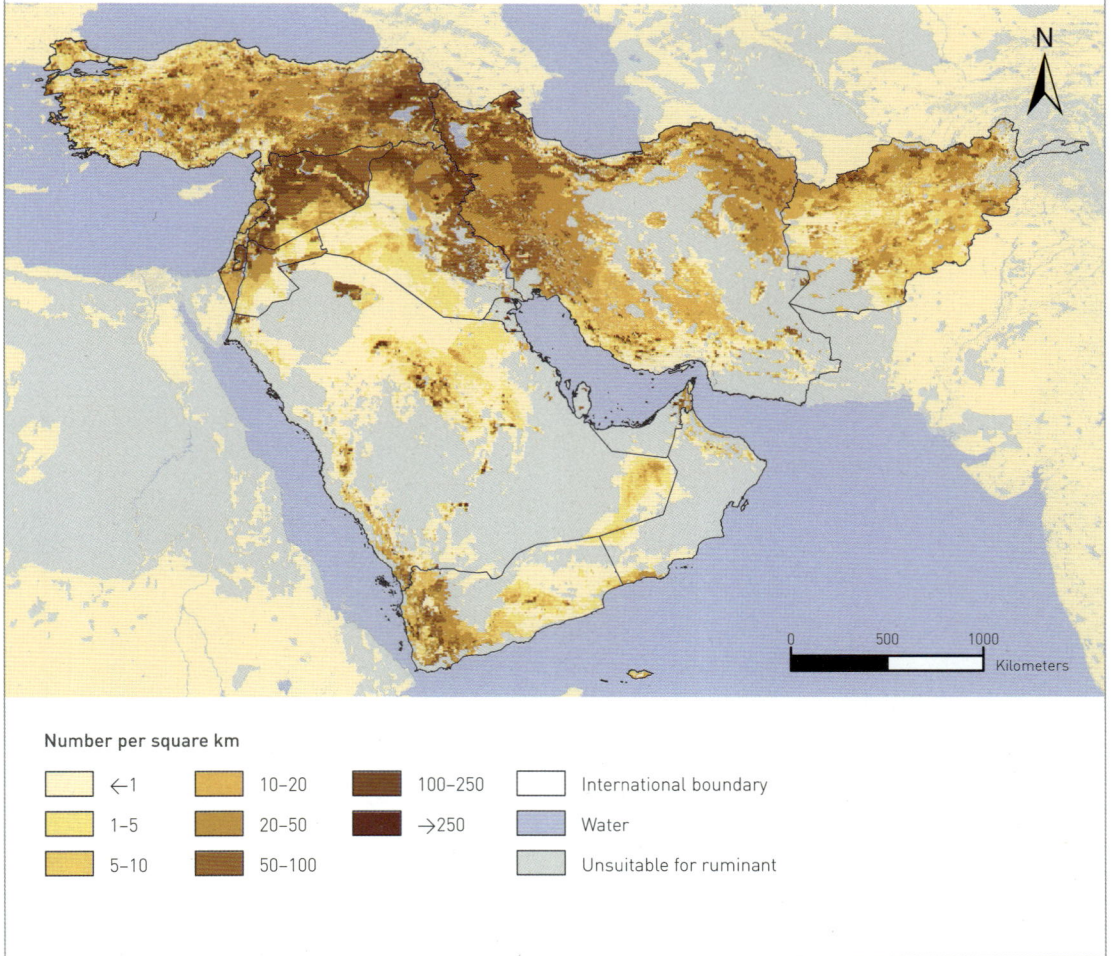

Number per square km

←1	10–20	100–250
1–5	20–50	→250
5–10	50–100	

International boundary
Water
Unsuitable for ruminant

6.8 DISTRIBUTION OF SHEEP IN AFRICA (Modelled)

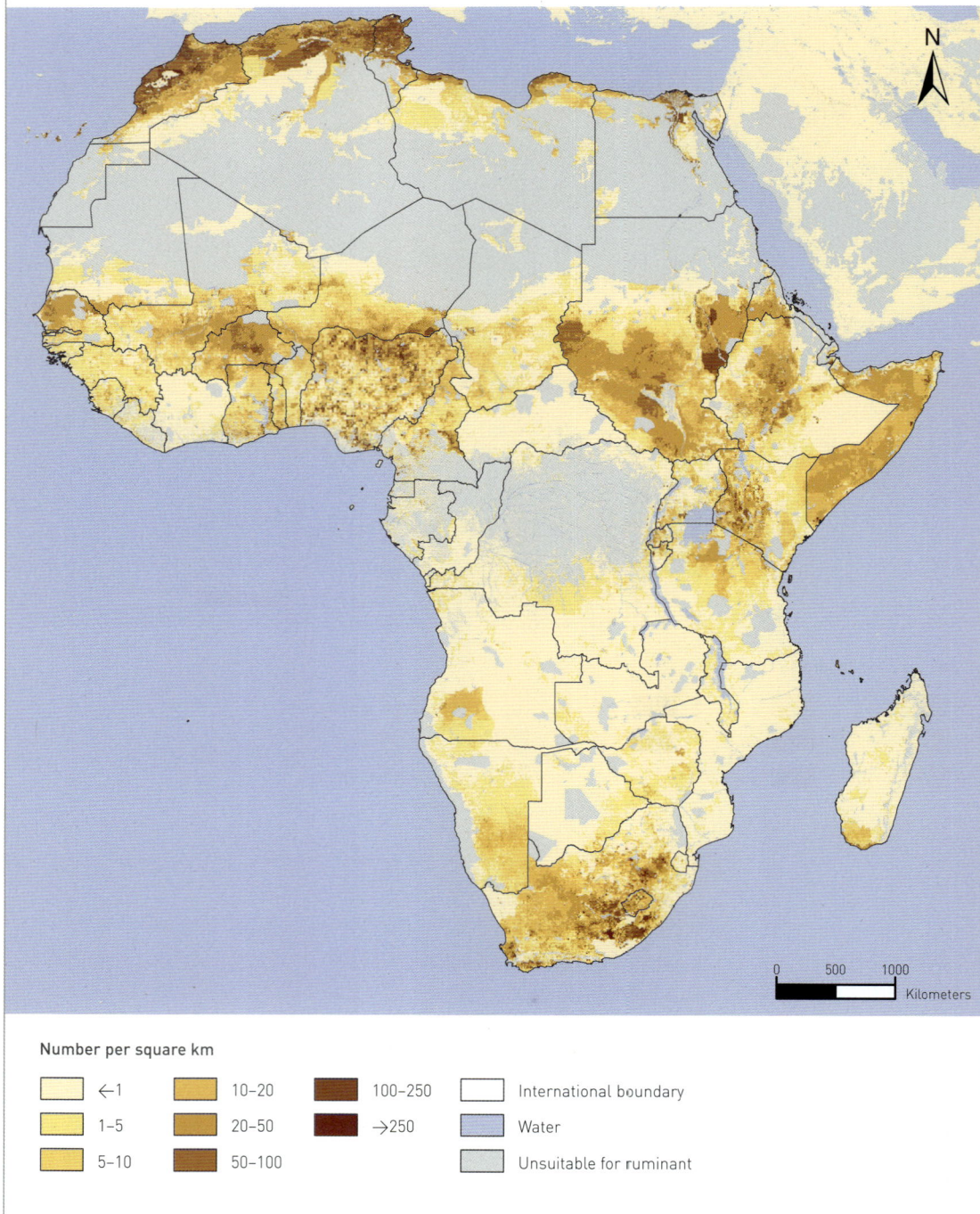

0   500   1000
Kilometers

Number per square km

←1	10–20	100–250
1–5	20–50	→250
5–10	50–100	

International boundary

Water

Unsuitable for ruminant

6.9 DISTRIBUTION OF GOATS IN THE FORMER SOVIET UNION (Modelled)

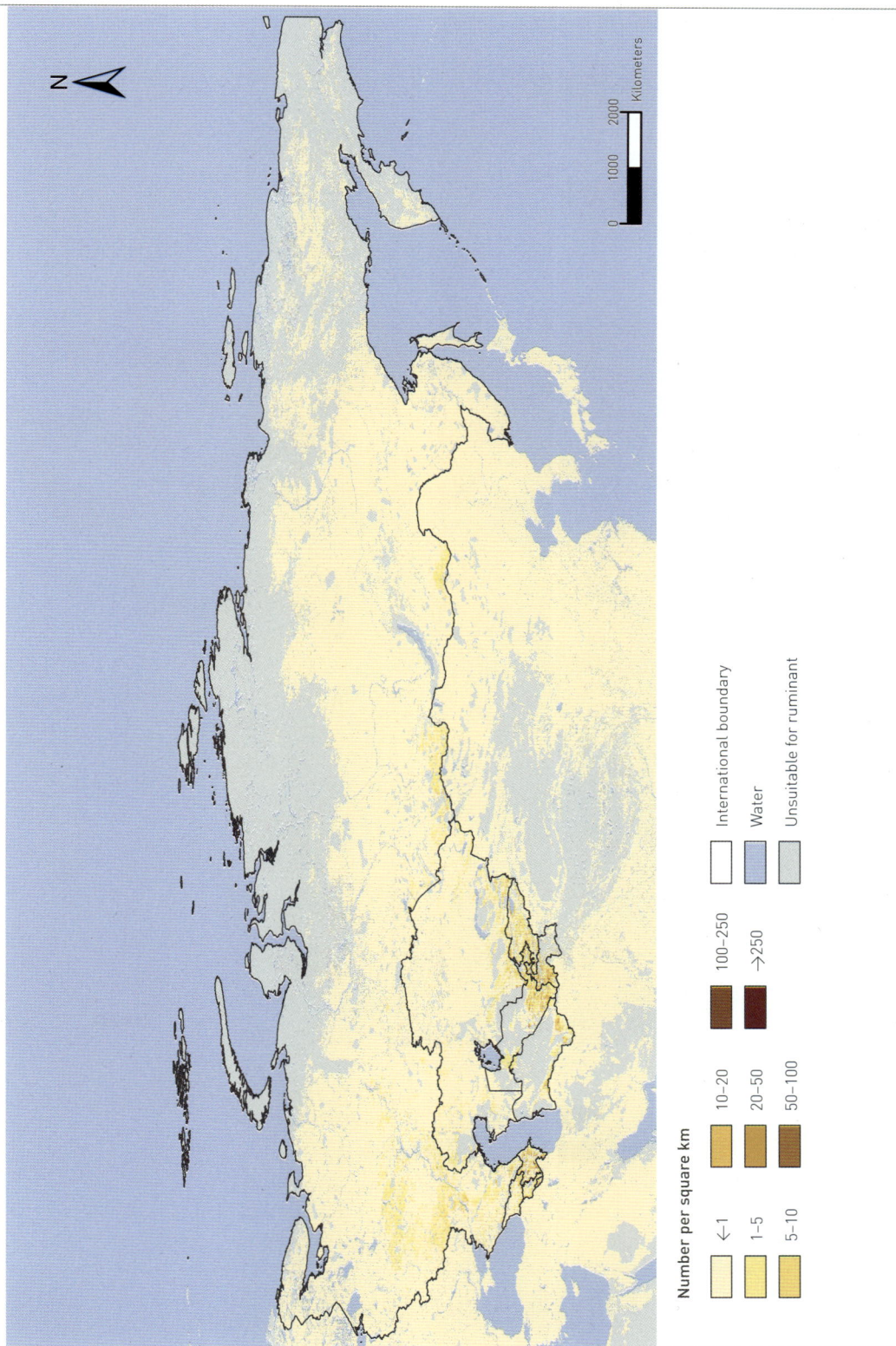

N

0 1000 2000
Kilometers

Number per square km

<1
1-5
5-10
10-20
20-50
50-100
100-250
→250

International boundary
Water
Unsuitable for ruminant

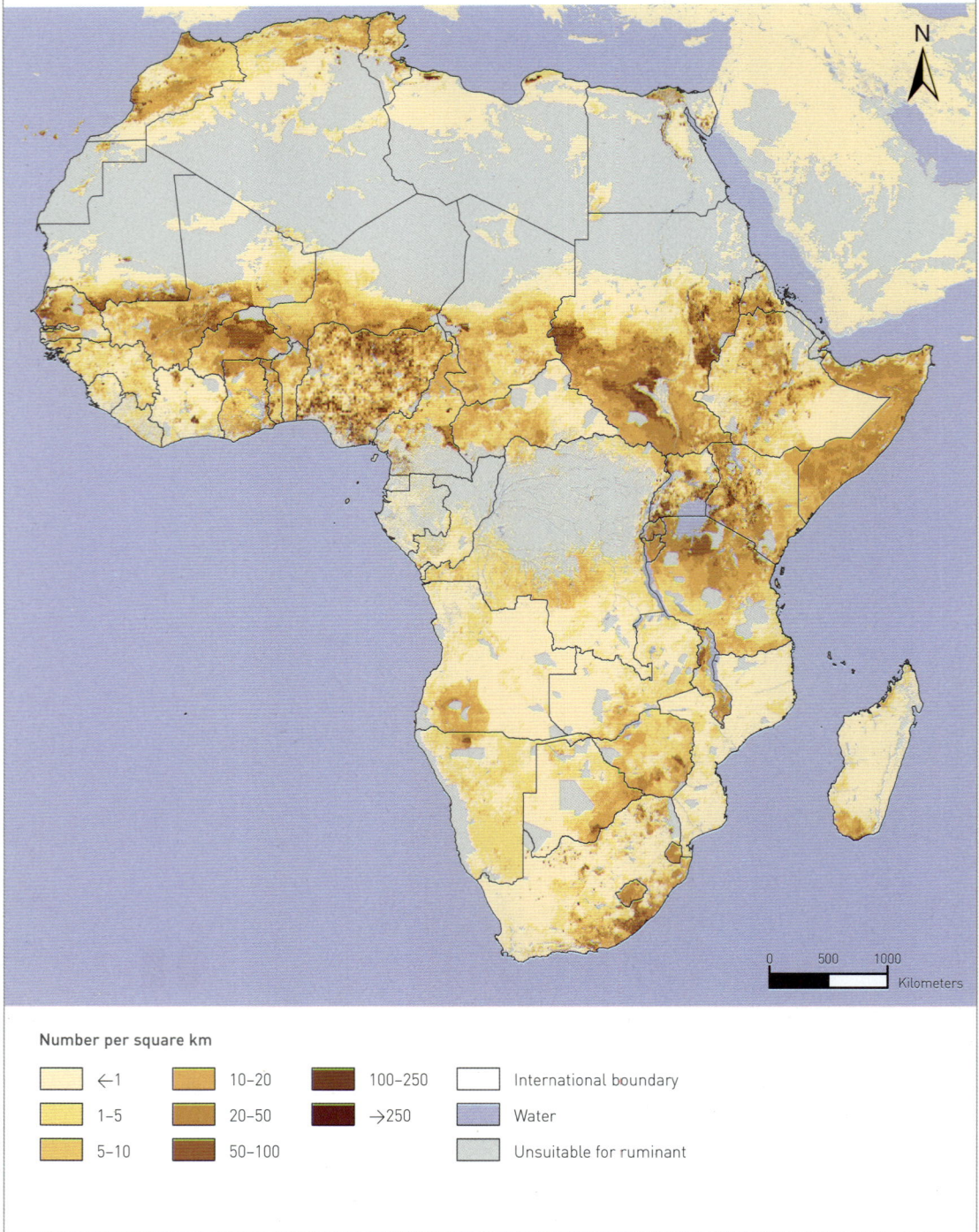

6.10 DISTRIBUTION OF GOATS IN AFRICA (Modelled)

Number per square km

- ←1
- 1–5
- 5–10
- 10–20
- 20–50
- 50–100
- 100–250
- →250
- International boundary
- Water
- Unsuitable for ruminant

6.11   GLOBAL PIG DISTRIBUTION (Modelled)

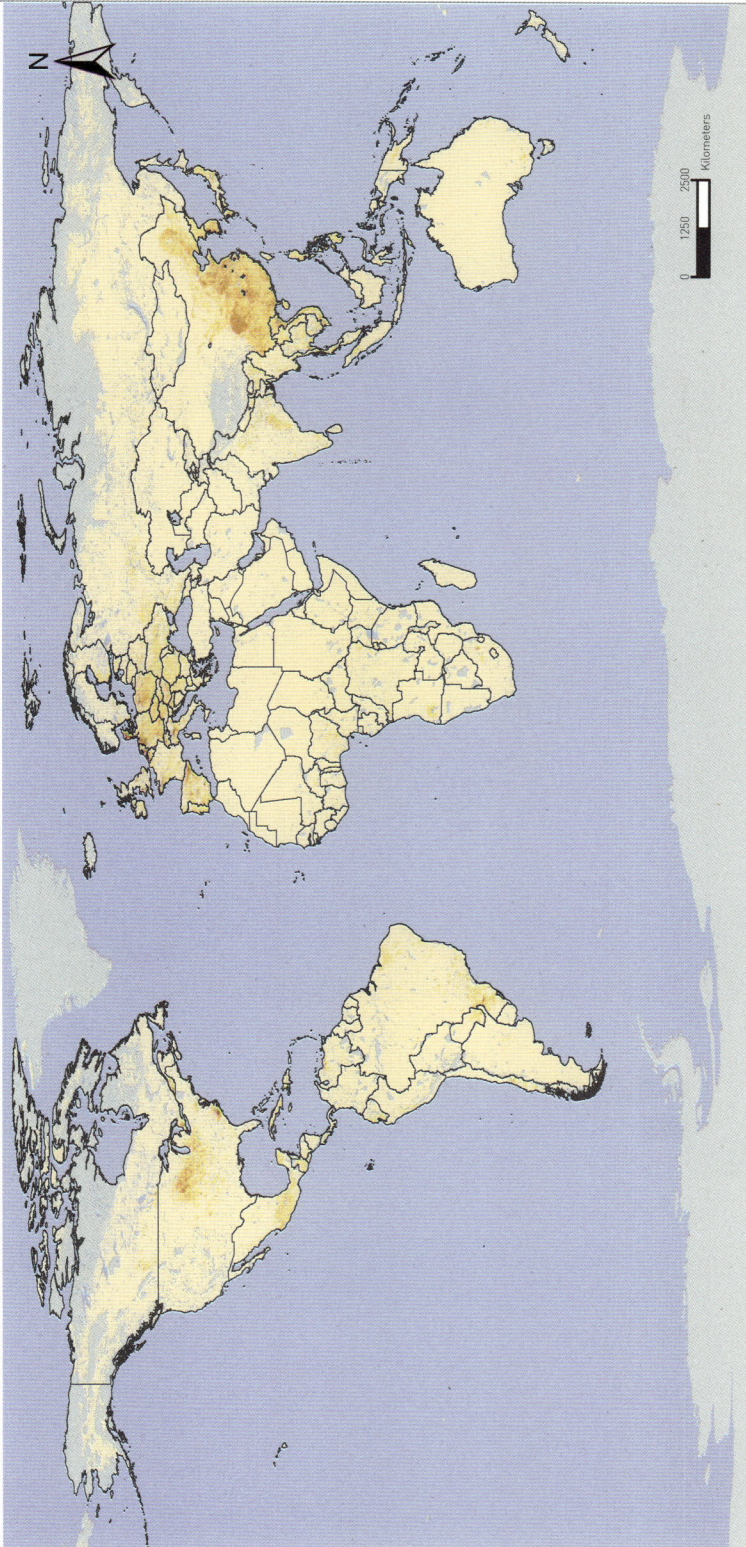

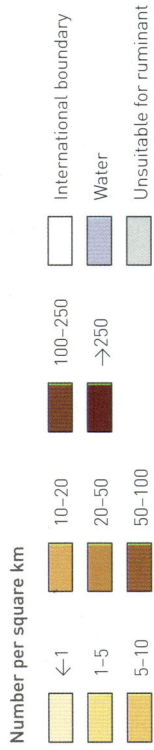

Number per square km

<1   10–20   100–250

1–5   20–50   >250

5–10   50–100

International boundary

Water

Unsuitable for ruminant

Kilometers
0   1250   2500

6.12   DISTRIBUTION OF PIGS IN EAST AND SOUTHEAST ASIA (Modelled)

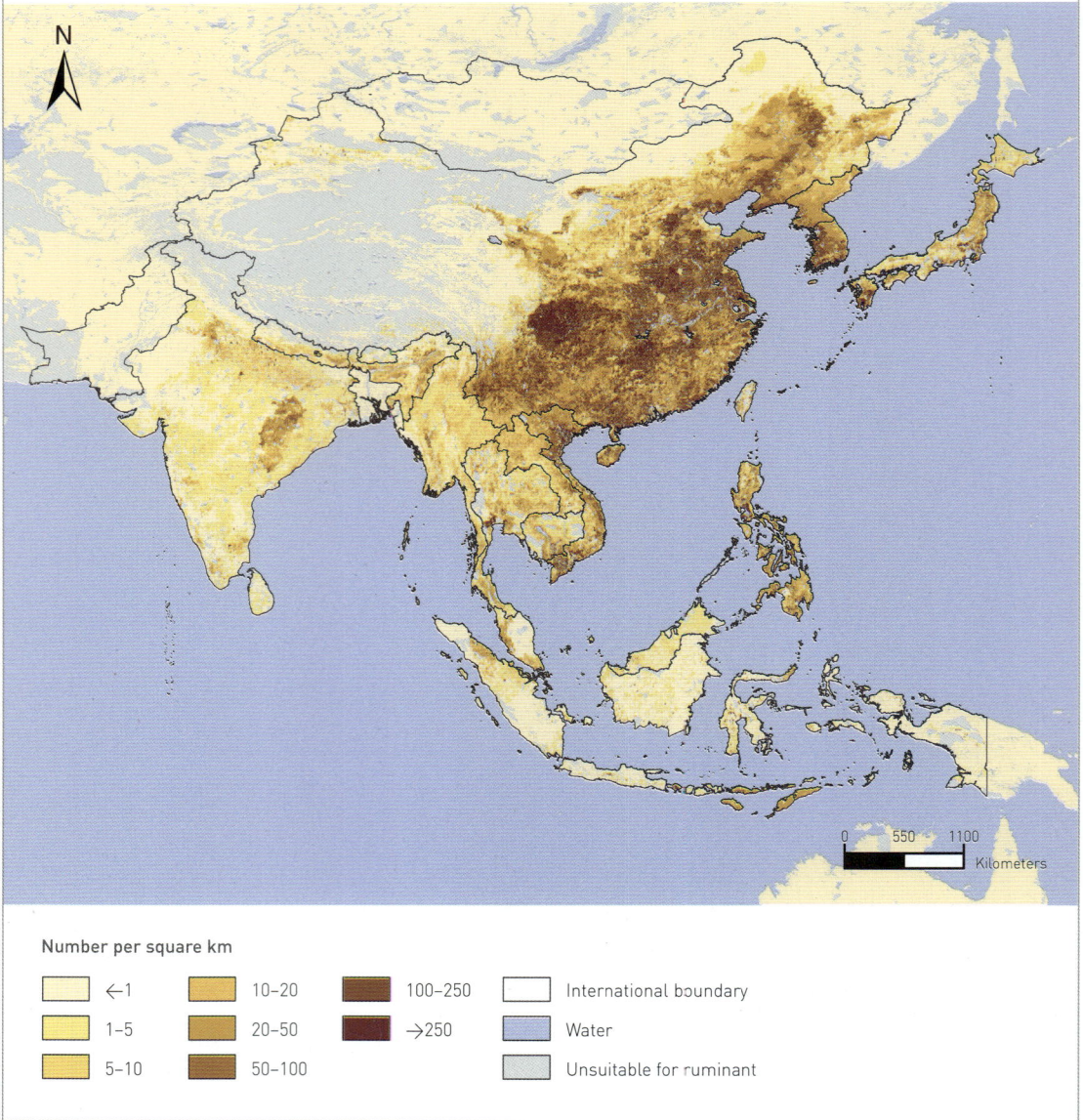

Number per square km

←1	10–20	100–250		International boundary
1–5	20–50	→250		Water
5–10	50–100			Unsuitable for ruminant

6.13 DISTRIBUTION OF PIGS IN NORTH AND CENTRAL AMERICA (Modelled)

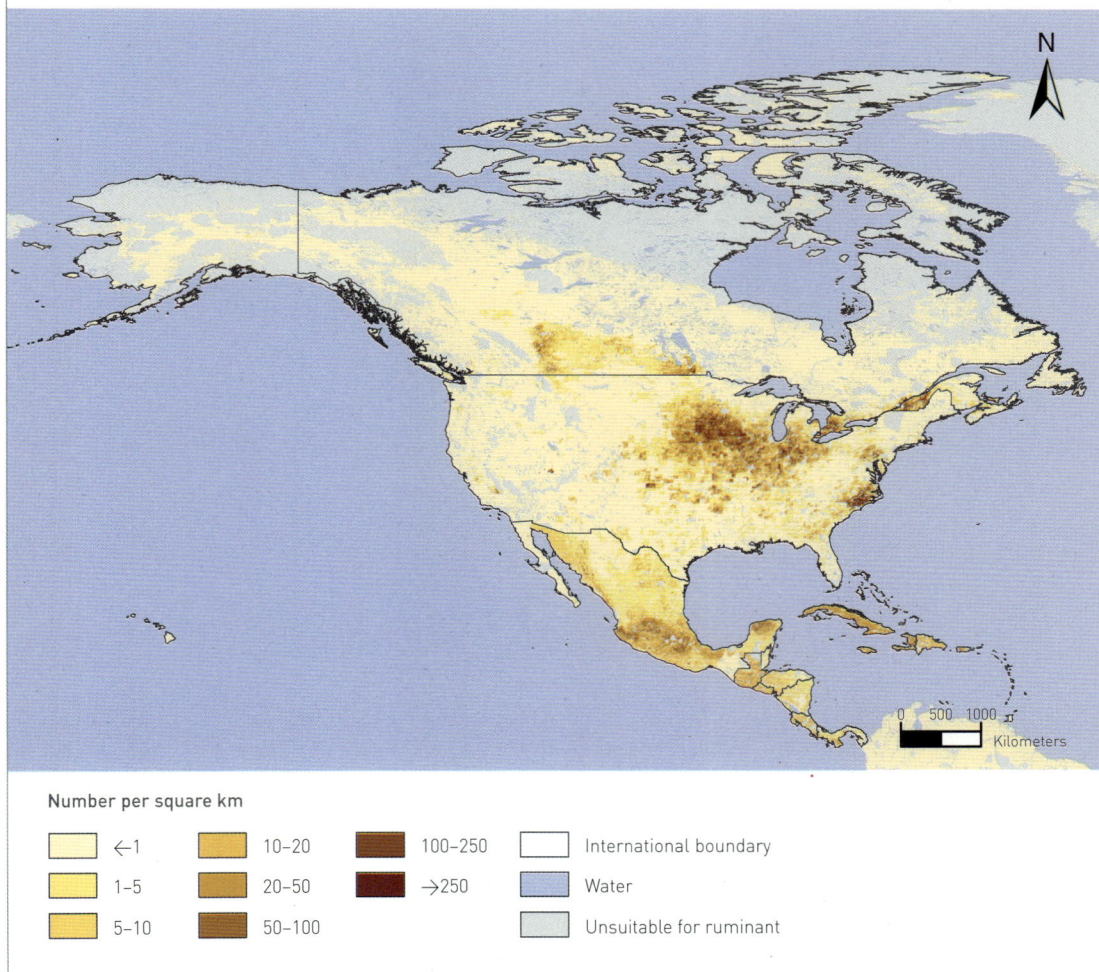

N

0   500   1000
Kilometers

Number per square km

←1	10–20	100–250	International boundary
1–5	20–50	→250	Water
5–10	50–100		Unsuitable for ruminant

6.14 GLOBAL POULTRY DISTRIBUTION (Modelled)

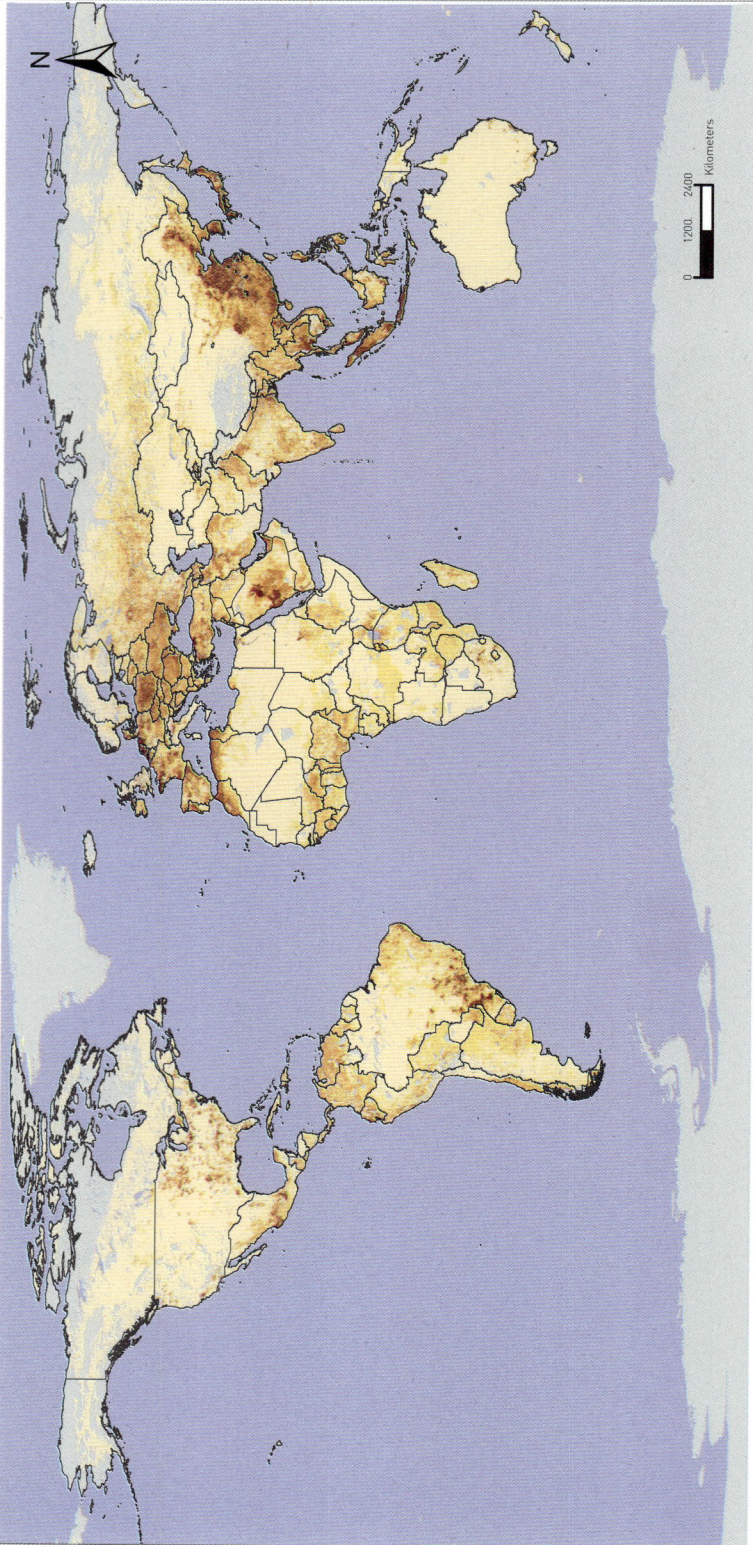

Number per square km

<1	10–20	100–250
1–5	20–50	→>250
5–10	50–100	

International boundary

Water

Unsuitable for ruminant

6.15 DISTRIBUTION OF POULTRY IN EAST AND SOUTHEAST ASIA (Modelled)

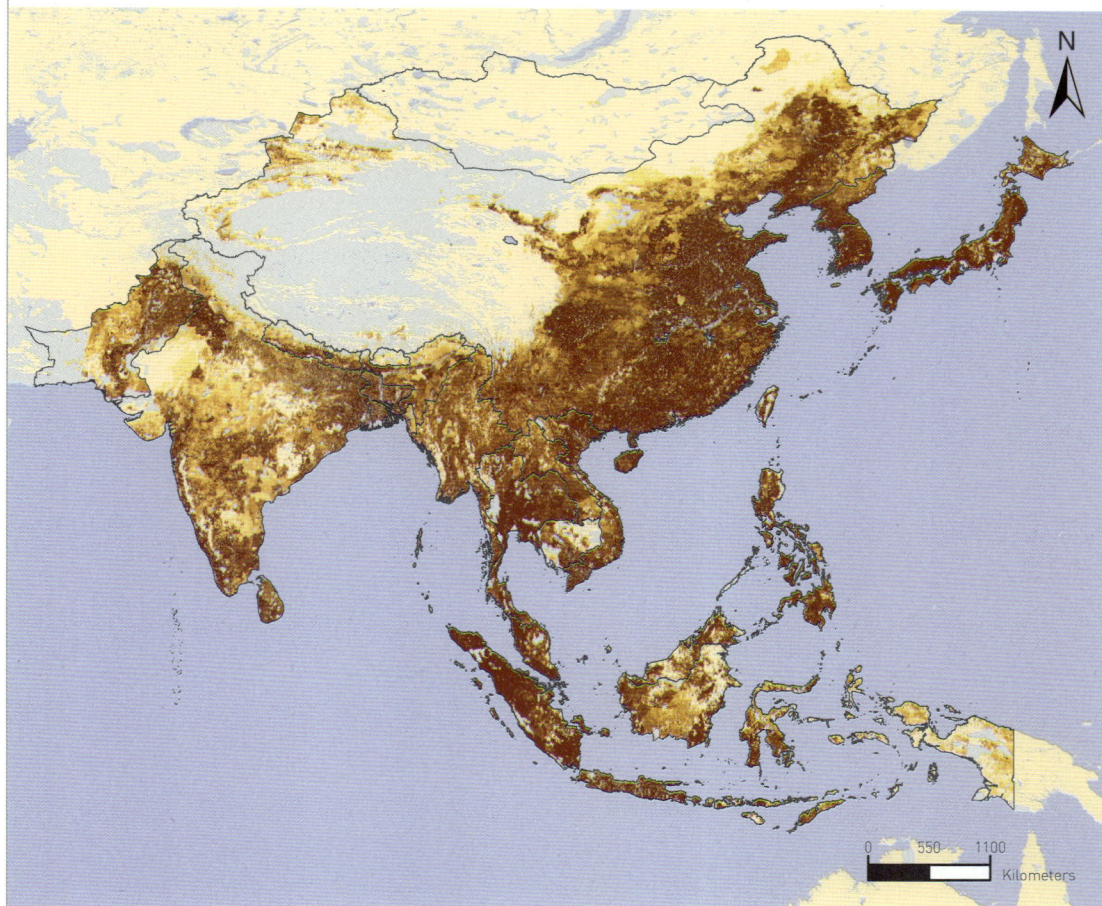

Number per square km

←1	10–20	100–250	International boundary
1–5	20–50	→250	Water
5–10	50–100		Unsuitable for ruminant

Source: Robinson *et al.*, 2007.

7 Applications

The spatial nature of these livestock data lends them uniquely to a wide array of applications. In essence, livestock distribution data provide the fundamental units for any analysis involving whole animals: for estimating production they provide the units to which production parameters may be applied; for evaluating impact (both of and on livestock), any number of different rates might be applied; and for epidemiological applications they provide the denominator in prevalence and incidence estimates, and the host distributions for transmission models. The range of potential applications of livestock distribution maps is boundless, but the following sections present just a few examples.

LIVESTOCK BIOMASS

Livestock populations are usually defined in terms of the number of individuals of a particular species in a given administrative region, or as standardized densities per unit area. The combination of individual species maps into an overall map of livestock distribution calls for the conversion of animal numbers into standard units of livestock biomass.

An example is given for the Mekong Region in Figure 7.1, wherein the distributions of cattle, buffaloes, small ruminants, pigs and chickens have been combined into a single map of livestock biomass measured in standard livestock units of 250 kg. In this case, livestock densities have been multiplied by animal live weights derived from FAO country-level estimates of carcass weights.

From maps such as these, the relative importance of monogastric species, for example, as opposed to ruminants, can be more confidently assessed. A single measure of livestock distribution also makes comparisons with other agricultural sectors and other regions easier.

Whilst one can envisage the value in combining ruminant species into a single composite value, for example, to estimate overall grazing pressure per unit of land, the value of combinations of species as disparate as cattle and chickens is less clear.

LIVESTOCK PROJECTIONS

The livestock distribution maps presented here are snapshots in time, although in reality livestock populations are not static. The most reliable way of assessing likely changes in livestock populations is to measure them through repeated surveys. However, given that such frequent data are rarely available, estimates need to be made. Projected changes in livestock population levels are regularly provided by FAO at the country level (see, for example, FAO, 2003). Whilst these values could be applied directly to modelled distributions, they would not reflect any change in the distribution of populations. To estimate re-distribution would either require the use of models of livestock spread (described below) or call for the linking of re-distribution to better-known parameters for which projections are available. In addition, given the close links between livestock distribution and environmental conditions, the potential effects of climate change should also be incorporated into medium- and long-term projections.

Some preliminary attempts have been made to project the spread of cattle in West Africa over a 20-year period as part of a study evaluating the economic impact of tsetse and trypanosomiasis control (Shaw *et al.*, 2006). These are described in the following subsections.

Carrying capacity and spread

The various elements of cattle population growth were calculated separately and then combined in several stages. First, breed-specific growth rates per animal, as supplied by herd growth models, were applied to a map of the current density of cattle to give first estimates of livestock growth.

7.1 DETAILED SPECIES MAPS FOR THE MEKONG REGION, COMBINED TO PRODUCE A MAP OF TLUs

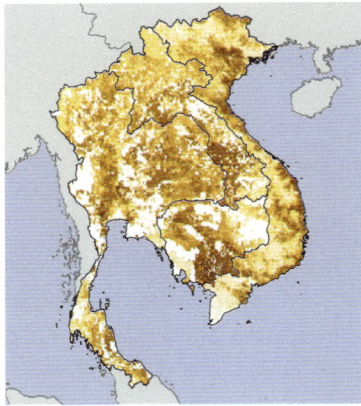

Cattle

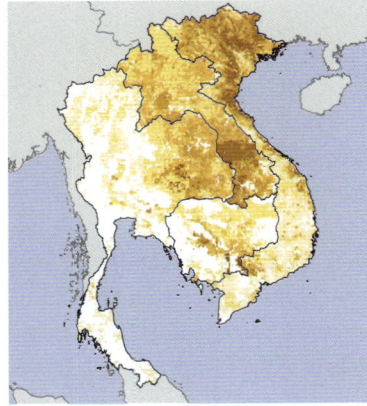

Buffaloes

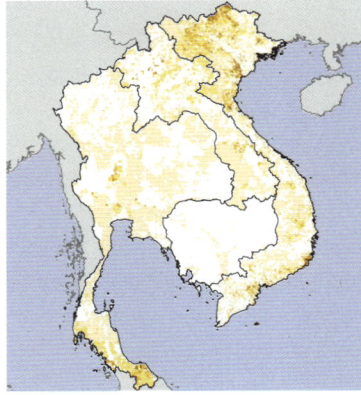

Small ruminants

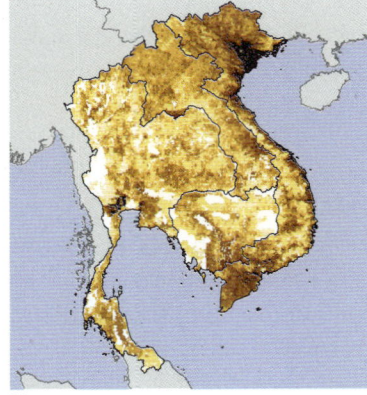

Pigs

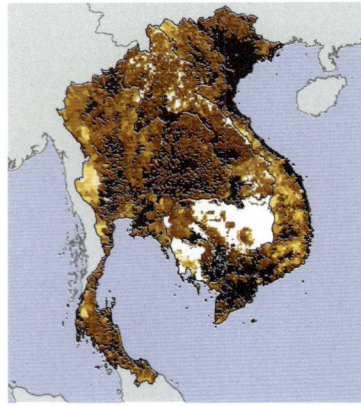

Chickens

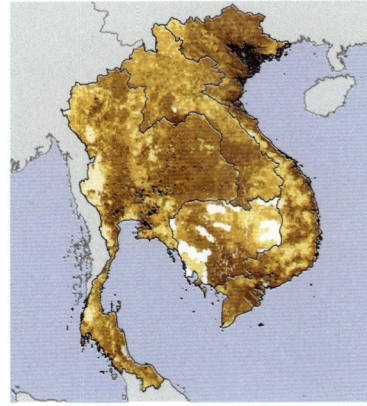

Tropical Livestock Units

Number per square km

0	1–5	10–20	50–100	→250
← 1	5–10	20–50	100–250	

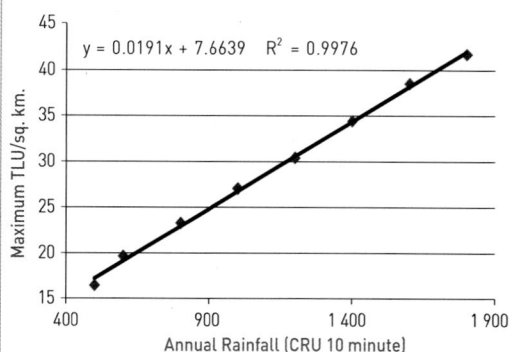

7.2 LIVESTOCK CARRYING CAPACITY AND ANNUAL RAINFALL

$y = 0.0191x + 7.6639 \quad R^2 = 0.9976$

Maximum TLU/sq. km.

Annual Rainfall (CRU 10 minute)

Source: Shaw et al., 2006, derived from Jahnke, 1982.

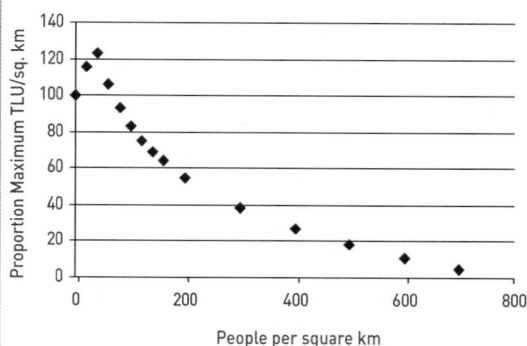

7.3 LIVESTOCK CARRYING CAPACITY AND HUMAN POPULATION

Proportion Maximum TLU/sq. km

People per square km

Source: Shaw et al., 2006.

When added to the existing population density, these provide an estimate of a theoretical cattle population after 20 years. This first output produces livestock population densities in some foci that significantly exceed likely carrying capacities, and must, therefore, be adjusted either by reducing calculated densities (equivalent to increasing off-take) or by 'exporting' animals from the high-concentration areas to surrounding, less heavily stocked regions.

The second of these possibilities has been adopted here, requiring first that carrying capacities are defined and, second, that techniques are developed to assign exported animals to neighbouring areas, as described in the following sections.

Mapping the carrying capacity

Carrying capacity is a controversial subject and, in recent years, the concept has fallen from favour amongst many ecologists. Nevertheless, livestock populations cannot increase indefinitely, and limits are reached beyond which animals are exported or slaughtered. Numerous attempts have been made to define thresholds for different zones (amongst which those cited in Jahnke, 1982), covering a range of rainfall bands. For the study area, these are summarized in Figure 7.2.

This relationship does not, however, incorporate

any influence of competing land use by cropping and/or human settlement, or the use of crop residues as fodder, or indeed the effects of mobile livestock populations in transhumant areas. Information on year-round carrying capacity in relation to human population density has been compiled by Shaw, 1986, based on work and studies originally reported in Putt et al., 1980, with values expressed as a proportion of the 'maximum' carrying capacity, with no human population, assumed here to be equivalent to that defined by Jahnke, 1982. The estimated relationship between livestock carrying capacity and human population density is shown in Figure 7.3.

For current purposes, these estimates, expressed in TLUs (where one TLU is equivalent to 250 kg of biomass) were converted to cattle densities (Figure 7.4) so as to match the units of the livestock density map (Figure 7.5). To do this, specific weights were assigned to types of cattle, as follows: 0.75 TLU for zebu cattle; 0.55 TLU for trypanotolerant taurine cattle; 0.705 TLU for low-productivity system oxen; and 0.74 TLU for high-productivity system oxen. It should be noted also that the estimated carrying capacity assumed that the land currently without cattle would be cleared or managed to make the habitat suitable for cattle keeping.

7.4 ESTIMATED CARRYING CAPACITY FOR CATTLE IN WEST AFRICA

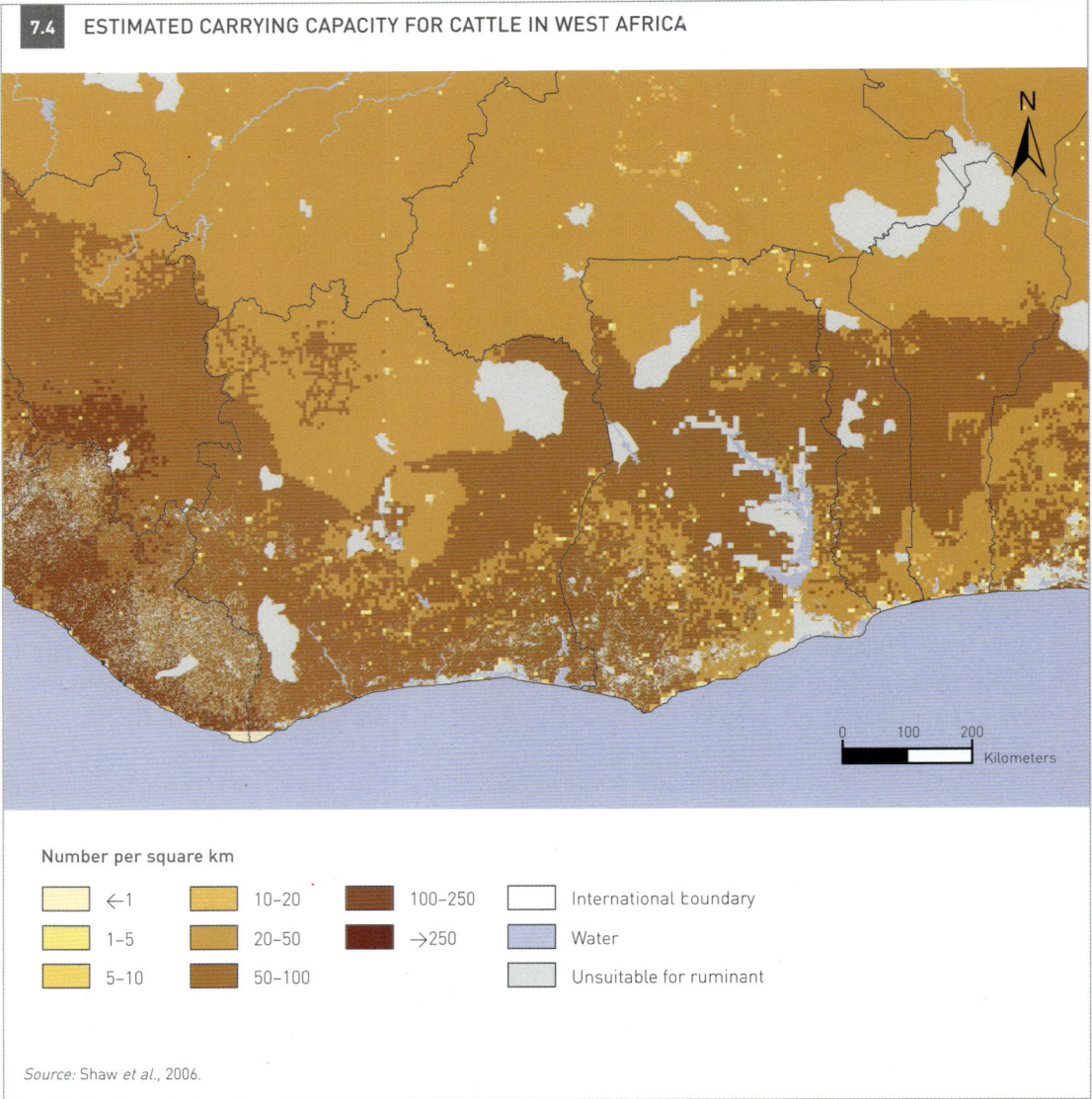

Number per square km

←1	10–20	100–250
1–5	20–50	→250
5–10	50–100	

International boundary
Water
Unsuitable for ruminant

Source: Shaw *et al.*, 2006.

Spread modelling

Methods of assigning emigrating populations to neighbouring areas from defined foci are still in their infancy. Some rely on simple diffusion, usually density-independent, and use some function of distance from the point of export to define areas of spread. Others attempt to incorporate the effect of long-distance dispersal events that emulate the establishment of new foci separated from the core areas: so-called 'stratified dispersal'. A recent set of models (Gilbert *et al.*, 2004) combines short- and long-range dispersal to define sequential areas of spread in 'time-steps', and allows for defining the rate of spread by short-range diffusion per time-step, as well as the number and maximum distance of new foci established over long distances. This is achieved by using the compound 'stratified' dispersal kernel shown as a red line in Figure 7.6, which combines the conventional short-distance curvilinear decrease (blue line) with a linear function to determine the probability of long-distance movements (black line), thereby increasing the

7.5 ESTIMATED EXISTING CATTLE DENSITIES IN WEST AFRICA

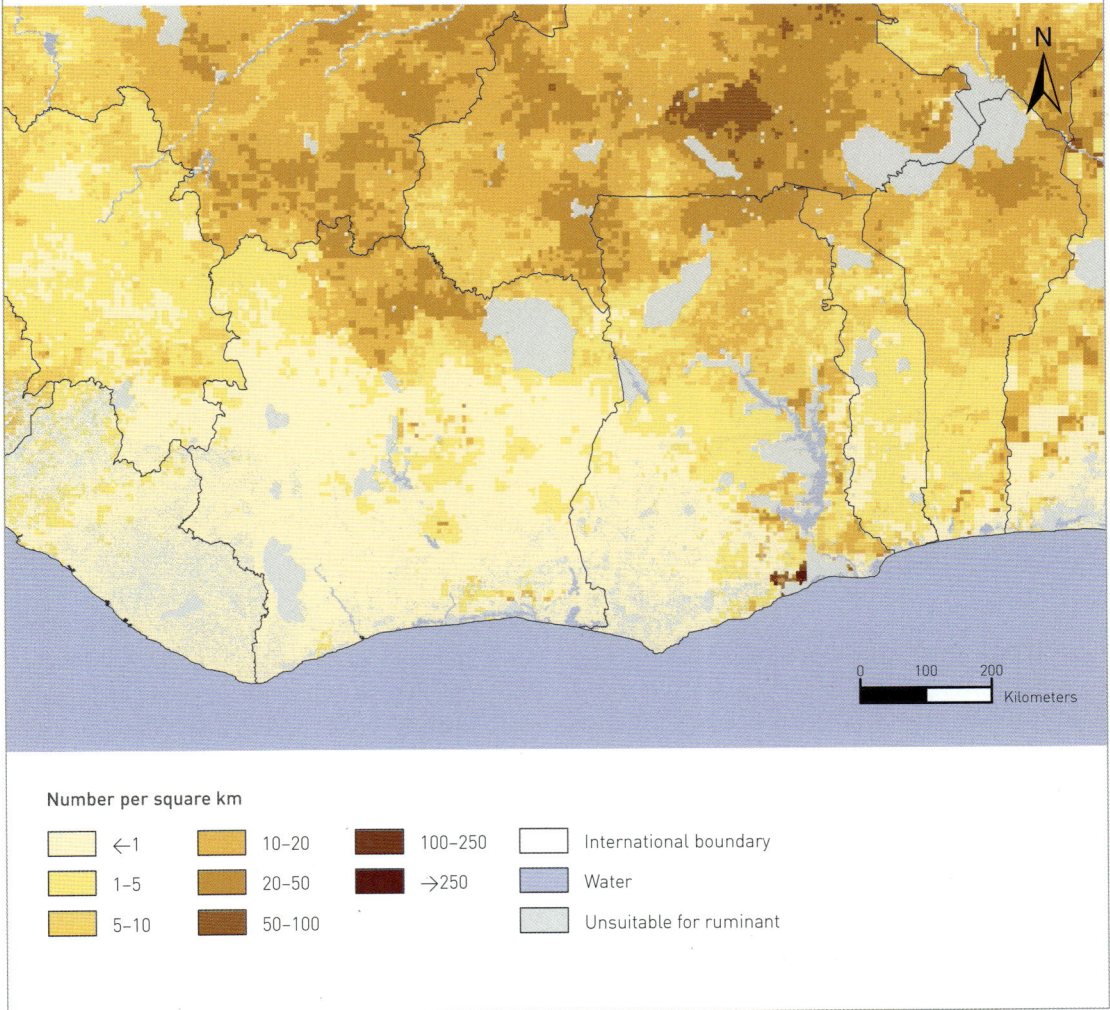

Number per square km

←1	10–20	100–250
1–5	20–50	→250
5–10	50–100	

International boundary

Water

Unsuitable for ruminant

numbers of long-distance establishment events without influencing the short-distance diffusion pattern.

This approach thus allows for the identification of sequential bands of expansion from known foci: in the current case, areas of overstocking. Each time-step is coded separately and therefore fixed proportions of the population to be exported can be assigned. In the analysis described here, four time-steps were defined and assigned 40 percent of the population to be exported from areas classified as overstocked to the first time-step; 30 percent to the

second; 20 percent to the third; and 10 percent to the fourth and final time-step. This means that 40 percent of the stock remained in the 'overstocked' areas, which assumes that some improved production system is adopted within 20 years. In each case, spread was prevented into areas defined as unsuitable for livestock and was scaled according to proximity to roads.

The resulting predicted livestock density after 20 years of tsetse and trypanosomiasis control is shown in Figure 7.7, for which the starting density was that given in Figure 7.5.

7.6 DISPERSAL KERNEL USED TO MODEL THE SPREAD OF LIVESTOCK

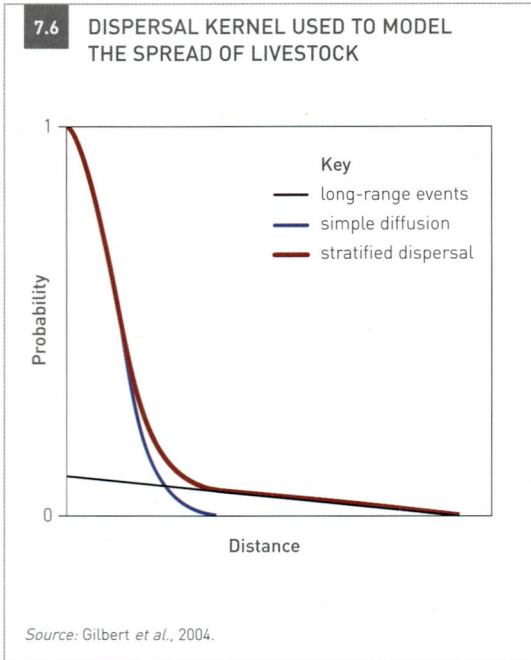

Source: Gilbert *et al.*, 2004.

LIVESTOCK PRODUCTION SYSTEM CLASSIFICATION

Livestock should not be considered in isolation from their surroundings, nor, as already illustrated in relation to biomass, should they be mapped only as single entities. The established links between livestock numbers, human populations and cultivation levels (Bourn and Wint, 1994) argue for paying greater attention to the quantification and mapping of these associations.

Since the 1970s, a number of farming system classifications have been proposed. Ruthenberg, 1980, for example, distinguishes among collection, cultivation and grassland utilization. For cultivation, his classification is based on the type and intensity of rotation used. For grassland utilization, Ruthenberg refers to the continuum from pure nomadism, through transhumance to sedentary animal husbandry.

Earlier, Grigg, 1972, had also distinguished characteristics of agriculture but failed to develop a systematic approach. This resulted in a rather disparate collection of systems and little reference to livestock production.

Seven broad farming systems mapped in a global study by the World Bank and FAO[39] combined current state-of-knowledge assessments of natural resources, prevailing farming activities and livelihood strategies to define them (Dixon *et al.*, 2001). This approach led to a classification based broadly on agro-ecology, presence or absence of irrigation and location (urban/coastal), but did not incorporate livestock in any detail.

Relatively simple statistical classifications of cattle and human population levels, cultivation intensity and elevation have also been investigated (Wint *et al.*, 1997; 1999). Whereas these classifications have the advantage of providing data-driven definitions of 'farming systems' and can delineate areas where these parameters have similar numerical values, they are sensitive both to geographical region and value range and cannot be replicated systematically in time and space.

FAO, 1996, developed a classification of livestock systems based on agro-ecology and the distinction between mixed and pastoral, irrigated and rainfed, and urban/landless areas. Emerging from this is one of the more widely used classifications developed and mapped by the International Livestock Research Institute (ILRI) (Thornton *et al.*, 2002). Figure 7.8 shows the decision tree that was used to map these livestock-oriented production systems.

The system is based on four modes of production (livestock grazing; rainfed crop and livestock production; irrigated crop and livestock production; and landless livestock production) in three agro-ecological zones defined by LGP and temperature (arid/semi-arid; humid/sub-humid; and temperate/tropical highlands). A number of global datasets was incorporated into the classification. The LGP (Fischer *et al.*, 2002)[40] was used to define all climatic zones except the highland temperate category, for which were used two climatic databases from the International Centre for Tropical Agriculture (Jones

[39] http://www.fao.org/farmingsystems/
[40] http://www.fao.org/waicent/faoinfo/agricult/agl/agll/gaez/index.htm

7.7 MODELLED CATTLE DENSITY AFTER 20 YEARS OF TSETSE AND TRYPANOSOMIASIS CONTROL IN WEST AFRICA

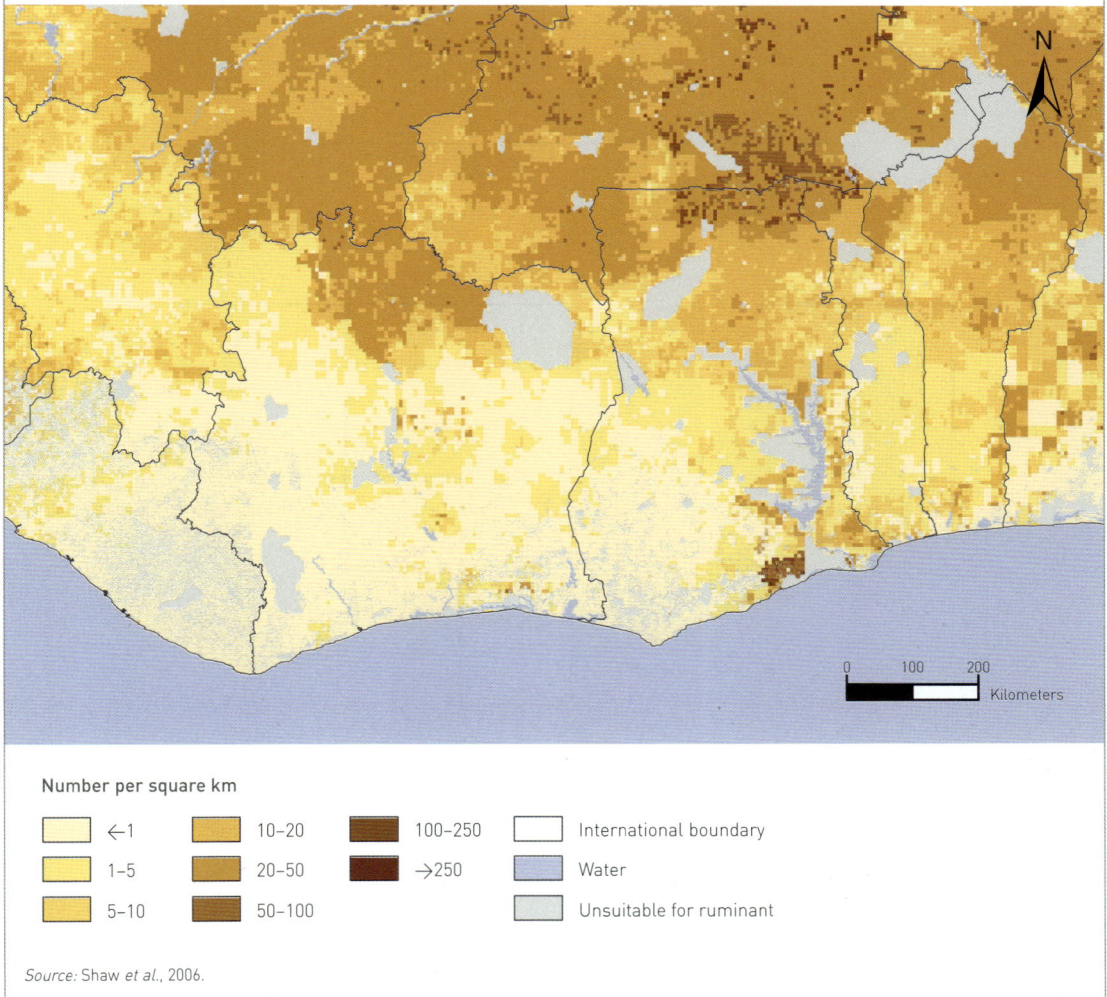

Number per square km

←1	10–20	100–250	International boundary
1–5	20–50	→250	Water
5–10	50–100		Unsuitable for ruminant

Source: Shaw *et al.*, 2006.

and Thornton, 1999) and the International Water Management Institute world water and climate atlas[41]. Irrigation data were taken from Doll and Seibert, 2000; elevation from the GTOPO30 DEM[42]; cropping was derived from a number of datasets, including those described in Loveland *et al.*, 2000; Anderson *et al.*, 1976, and Wood *et al.*, 2000, but was heavily dependent both on interpretation and on expert opinion; night-time lights were taken

from the NOAA/National Geophysical Data Center Stable Lights and Radiance Calibrated Lights of the World[43]; and human population data were taken from various sources (Hyman *et al.*, 2000; Reid *et al.*, 2000; Deichmann, 1996)[44]. Figure 7.9 provides an example of the resultant livestock production systems classification for Africa.

The above livestock production system classification does not incorporate livestock population data

[41] http://www.iwmi.cgiar.org/watlas/atlas.htm
[42] http://edc.usgs.gov/products/elevation/gtopo30/gtopo30.html

[43] http://www.ngdc.noaa.gov/dmsp/download_night_time_lights_94-95.html
[44] http://grid2.cr.usgs.gov/globalpop/

7.8 **DECISION TREE FOR MAPPING LIVESTOCK PRODUCTION SYSTEMS**

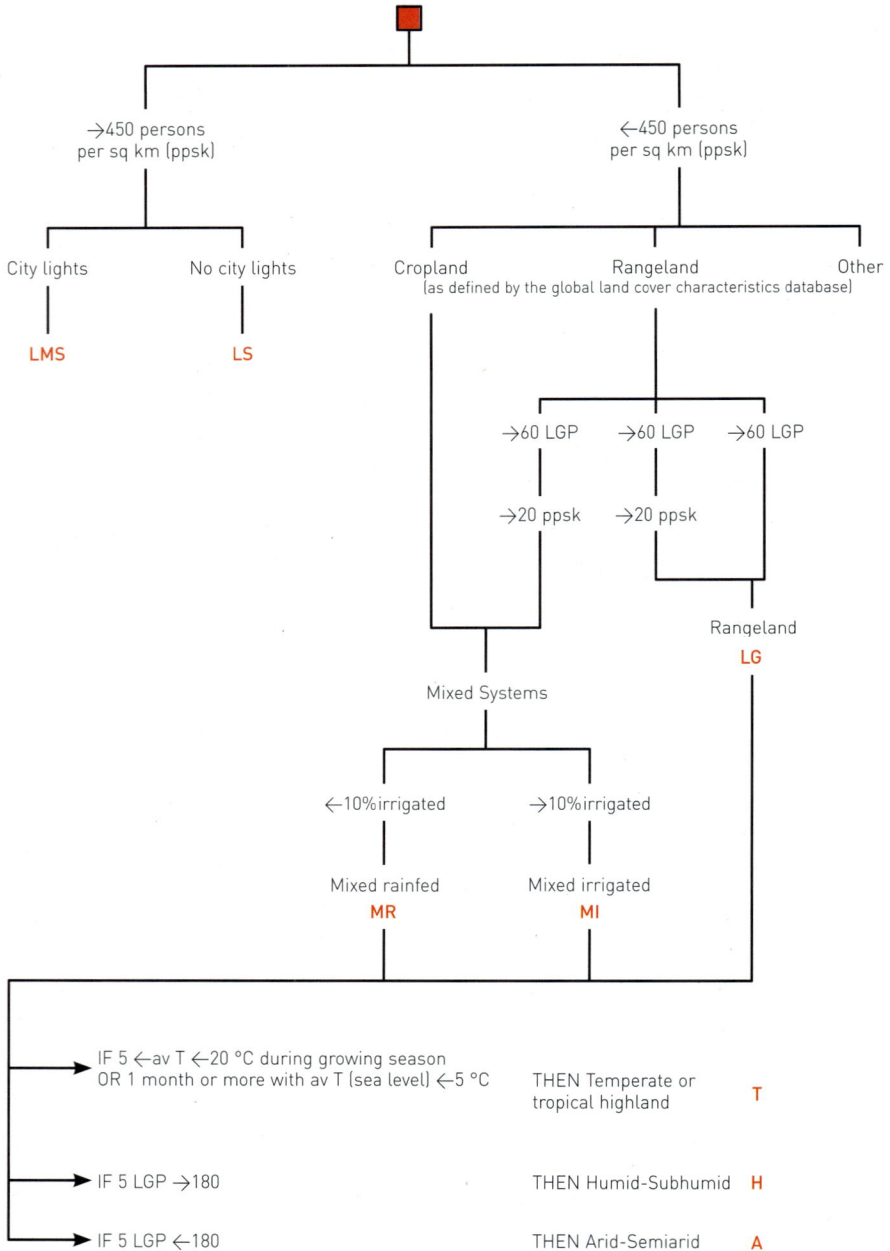

→450 persons per sq km (ppsk)

←450 persons per sq km (ppsk)

City lights

No city lights

Cropland
(as defined by the global land cover characteristics database)

Rangeland

Other

LMS

LS

→60 LGP

→60 LGP

→60 LGP

→20 ppsk

→20 ppsk

Rangeland
LG

Mixed Systems

←10% irrigated

→10% irrigated

Mixed rainfed
MR

Mixed irrigated
MI

IF 5 ←av T ←20 °C during growing season
OR 1 month or more with av T (sea level) ←5 °C

THEN Temperate or
tropical highland

T

IF 5 LGP →180

THEN Humid-Subhumid

H

IF 5 LGP ←180

THEN Arid-Semiarid

A

Source: Thornton *et al.*, 2002.

7.9 LIVESTOCK PRODUCTION SYSTEMS IN AFRICA

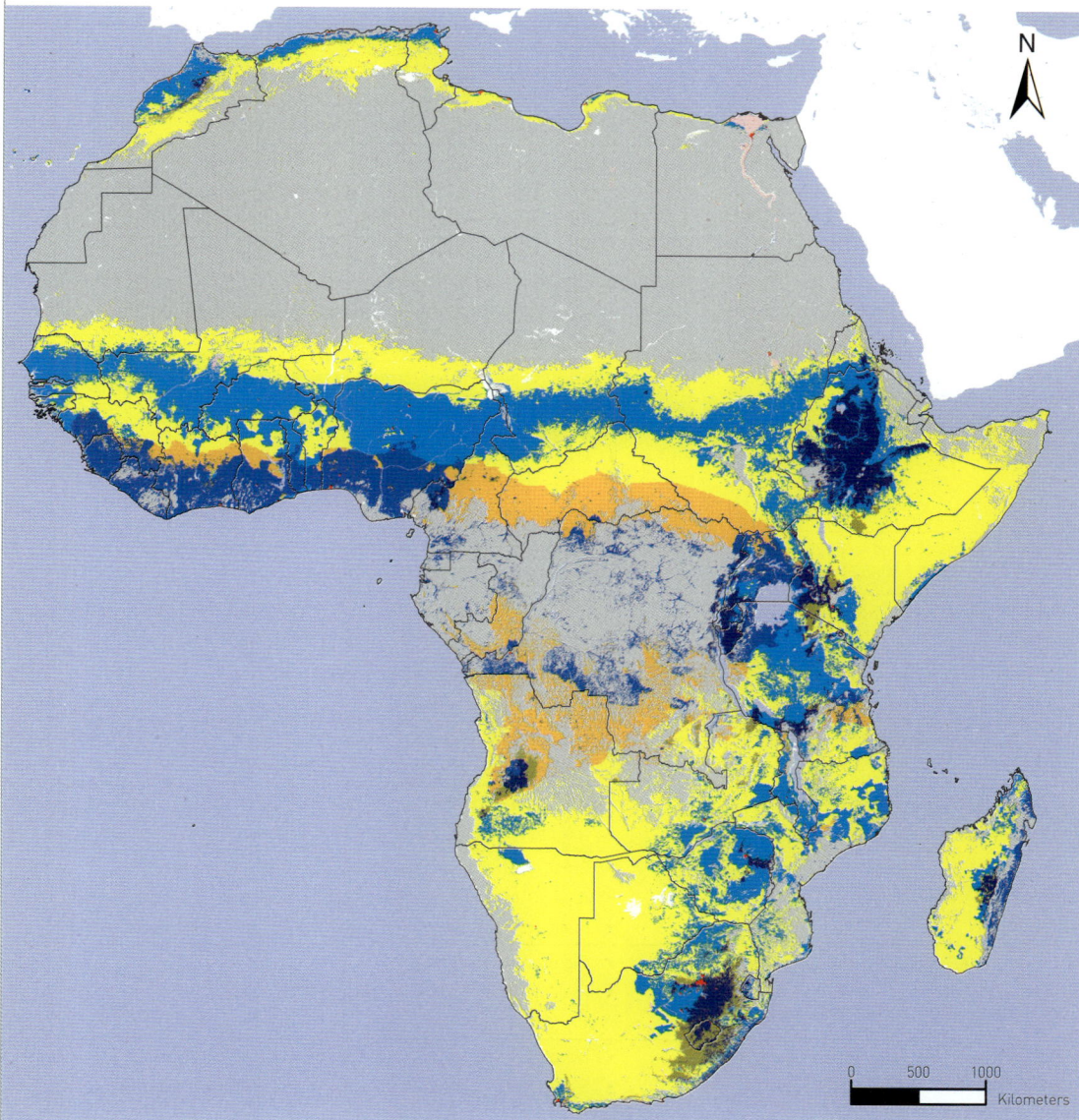

Production systems

■ LGA	■ MIA	■ MRA	■ Other	□ International boundary
■ LGH	■ MIH	■ MRH	■ Urban	■ Water
■ LGT	■ MIT	■ MRT		

Source: Reproduced from Thornton *et al.*, 2002.

TABLE 7.1 DESCRIPTIONS AND EXAMPLES OF LIVESTOCK PRODUCTION SYSTEMS

Production system	Examples
LGT: Temperate and tropical highlands (COLD GRASS)	• Mongolia's steppe system • Dairy systems near Bogota, Colombia; Peru and Bolivia Altiplano camelid and sheep-grazing systems • Chinese merino wool sheep on communal grazing
LGH: Humid/subhumid tropics and subtropics (WET GRASS)	• Extensive ranching in South American lowlands • Ranching systems in West and Central Africa • Amazonian ranching
LGA: Arid/semi-arid tropics and subtropics (DRY GRASS)	• Pastoralists in the Sahel • Near East and North Africa pastoralists • Beef-milk systems on pastures in Mexico, Venezuela • Southern Africa ranches
MRT: Temperate and tropical highlands mixed rainfed (COLD MIXED)	• Smallholder peasant farmers in northern China • Smallholders in Ethiopian highlands where oxen for traction is important • Mixed crop-livestock smallholders in highlands of Central and South America • Smallscale peri-urban dairy in East African highlands
MRH: Humid/subhumid tropics and subtropics mixed rainfed (WET MIXED)	• Areas of South America where rainforests are being cleared • Large areas of sub-Saharan Africa (tsetse 'belt')
MRA: Arid/semi-arid tropics and subtropics (DRY MIXED)	• Dryland farming-sheep systems in West Asia-North Africa and India • Small ruminant-cassava systems in northeast Brazil • Mixed crop-livestock farms in Burkina Faso, Nigeria • Dairy in Senegal and Mali
MIT: Temperate and tropical highlands mixed irrigated (COLD IRRIGATED)	• Mediterranean region • Far East Asian irrigated rice/dairy farms
MIH: Humid/subhumid tropics and subtropics mixed irrigated (WET IRRIGATED)	• Irrigated rice-buffalo systems of the Philippines, Viet Nam and India • Irrigated rice, pig and poultry enterprises in Asia
MIA: Arid/semi-arid tropics and subtropics mixed irrigated (DRY IRRIGATED)	• Small-scale buffalo milk production, Pakistan & India • Animal-traction-based cash crop production in Egypt and Afghanistan • Intensive dairy systems in California (United States), Israel, Mexico
Other/Urban: Landless mono-gastric systems: value of production of the pig/poultry enterprises → the ruminant enterprises	• Pig production in Asia • Poultry production in Central and South America
Other/Urban: Landless ruminant systems: value of production of the ruminant enterprises → the pig/poultry enterprises	• Landless sheep production systems in West Asia-North Africa • Sheep fattening operations in Syria and Nigeria

Source: Reproduced from Thornton *et al.*, 2002.

in its definition. It tends to amalgamate similar systems and fails to capture important differences in use and livestock husbandry practices within categories, e.g. grassland-based grazing combines pastoralists and ranchers, which are clearly not equivalent. However, it is undoubtedly the most appropriate classification system available, and does provide a relevant stratification through which to describe, visualize and explore livestock and livestock-related issues. Table 7.1, reproduced from Thornton *et al.*, 2002, gives examples of systems around the world that fall under each of the 11 categories defined.

When these production systems are combined with the gridded livestock data presented here, estimated numbers of livestock can be extracted by production system. Compared with simple national totals, this gives a more meaningful breakdown of how livestock are distributed across the globe. As an example that shall be returned to later, Table 7.2 gives a breakdown of cattle numbers (in this case adjusted to FAOSTAT 2005 national totals) by livestock production system in the Horn of Africa. Tables of global livestock numbers by country and, where available, by production system, are given in the annex hereto.

In their original application of these livestock production systems, Thornton *et al.*, 2002, used them to delineate and extract a number of socio-economic variables. They produced tables, for each production system in developing countries, of estimates of the numbers of people, poor people and poor livestock keepers. This type of application is useful for regional targeting and for impact assessment. Since the systems are defined in terms of population density and LGP, the classification can be re-evaluated using different scenarios of population and LGP. A tentative assessment of how these systems might be transformed by human population growth and climate change was thus made, giving some clues as to how the distribution of farming systems may change in the future.

The original livestock production system maps produced by Thornton *et al.*, 2002, did not provide global coverage. However, collaborative work is ongoing at FAO and ILRI to further develop and standardize global livestock production system maps for a number of applications.

LIVESTOCK PRODUCTION ESTIMATES

Livestock production and off-take rates vary across different livestock production systems, and in a broadly predictable way. This introduces a further application that involves livestock production sys-

tem maps to provide a stratification scheme within which to parameterize livestock growth and off-take models.

The GLW maps of livestock densities have been used to map production and off-take levels in sub-Saharan Africa (Otte *et al.*, 2001). For example, beef and milk production and use of draught power per square kilometre have been estimated by deriving annual output per head of cattle within each of seven major agro-ecological zones. These zones were defined and mapped by combining a number of spatial variables (temperature, elevation, LGP and crop type) in a decision tree (FAO, 2002b); livestock production was modelled for each zone using the herd growth model within the Livestock Development Planning System Version 2 (LDPS-2) (FAO, 1997). The herd models were parameterized separately for each zone, based on available published data (for some parameters, data were sparse). These production maps can then be further combined with human population density maps to produce estimates of off-take per capita.

To illustrate the above, in Figures 7.10 and 7.11, meat and milk off-take has been re-evaluated using both the new GLW grids for Africa (Figure 5.4) and the Thornton *et al.*, 2002, livestock production systems (Figure 7.9) to stratify production modelling.

These production maps of Africa are part of an ongoing FAO effort to map livestock production globally, based on the GLW datasets.

LIVESTOCK PRODUCTION BALANCE

Detailed Information on levels of international trade in livestock products is sparse, most especially that which provides consistent and complete global coverage. The situation is improving, however, and the Commodities and Trade Division of FAO has compiled estimates of imports and exports of livestock products at the country level, largely derived from FAOSTAT data[45]. In some cases, details are even provided on the countries to which a specific

[45] http://faostat.fao.org/

TABLE 7.2 CATTLE NUMBERS BY COUNTRY AND LIVESTOCK PRODUCTION SYSTEM IN THE HORN OF AFRICA, ADJUSTED TO FAOSTAT 2005
NATIONAL TOTALS

Production System	Djibouti	Ethiopia	Eritrea	Kenya	Somalia	Sudan	Uganda	System Total
LGA	137 070	3 743 525	861 950	2 932 925	4 264 535	16 443 100	262 500	28 645 605
LGH	n.a.	43 625	n.a.	n.a.	n.a.	967 500	208 570	1 219 695
LGT	n.a.	220 300	21 960	512 775	1 380	6 320	10 560	773 295
MIA	n.a.	7 150	610	n.a.	79 315	400 150	n a	487 225
MRA	8 760	8 735 260	693 880	2 044 045	372 780	17 751 500	1 536 415	31 142 640
MRH	n.a.	1 280 250	n.a.	1 006 430	n.a.	8 230	2 791 160	5 086 070
MRT	n.a.	23 198 000	175 580	4 030 505	n.a.	20 000	856 155	28 280 240
Urban	0	0	n.a.	13 470	0	9 000	0	22 470
Other	151 170	1 271 890	196 020	1 459 850	631 990	2 719 200	434 640	6 864 760
Country Total	297 000	38 500 000	1 950 000	12 000 000	5 350 000	38 325 000	6 100 000	102 522 000

Notes: Livestock production system data were taken from Thornton *et al.,* 2002.
'n.a.' indicates that system does not occur in a country.

7.10 ESTIMATED MEAT OFF-TAKE FROM CATTLE IN AFRICA

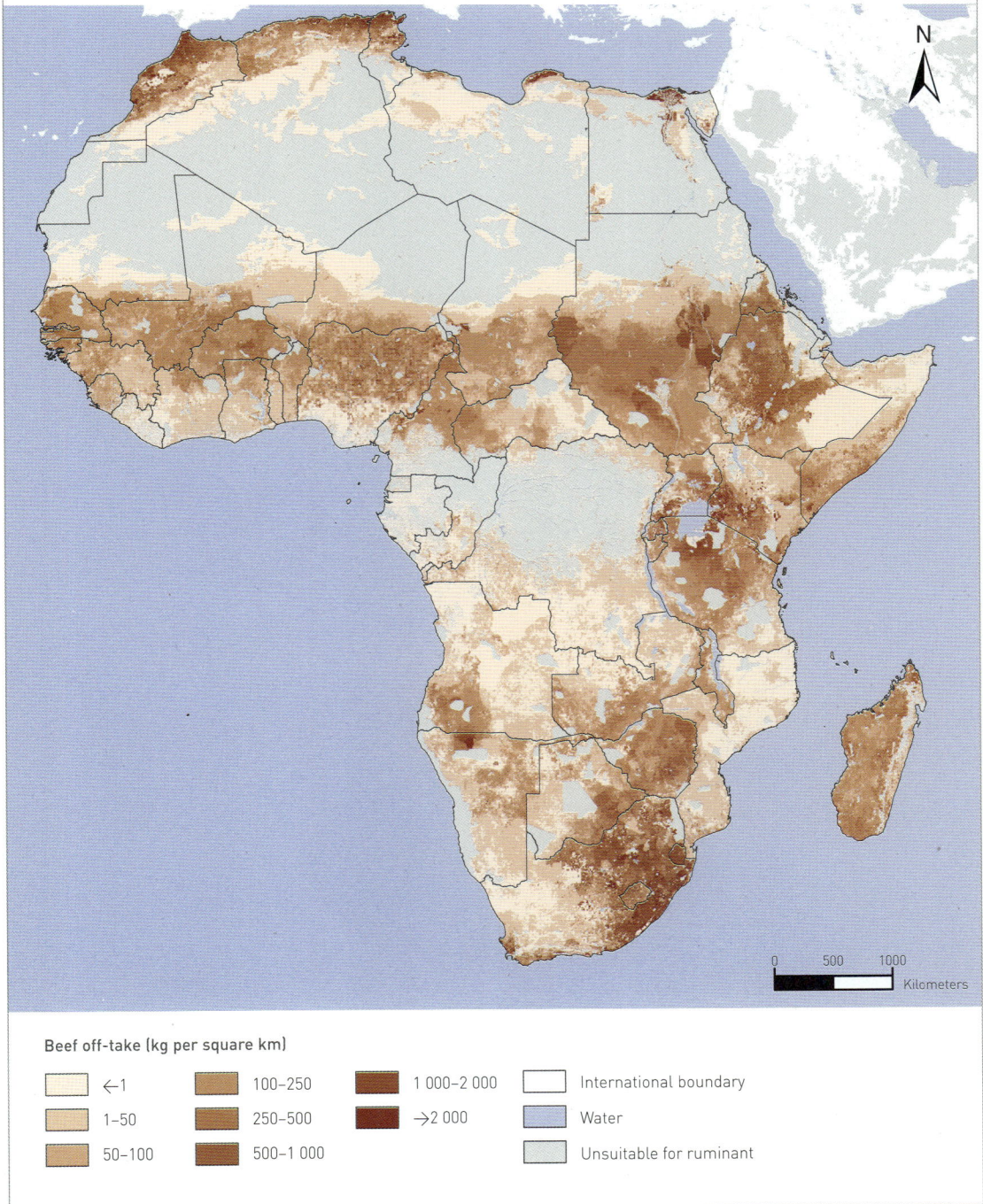

Beef off-take (kg per square km)

←1	100–250	1 000–2 000
1–50	250–500	→2 000
50–100	500–1 000	

International boundary

Water

Unsuitable for ruminant

7.11 ESTIMATED MILK OFF-TAKE FROM CATTLE IN AFRICA

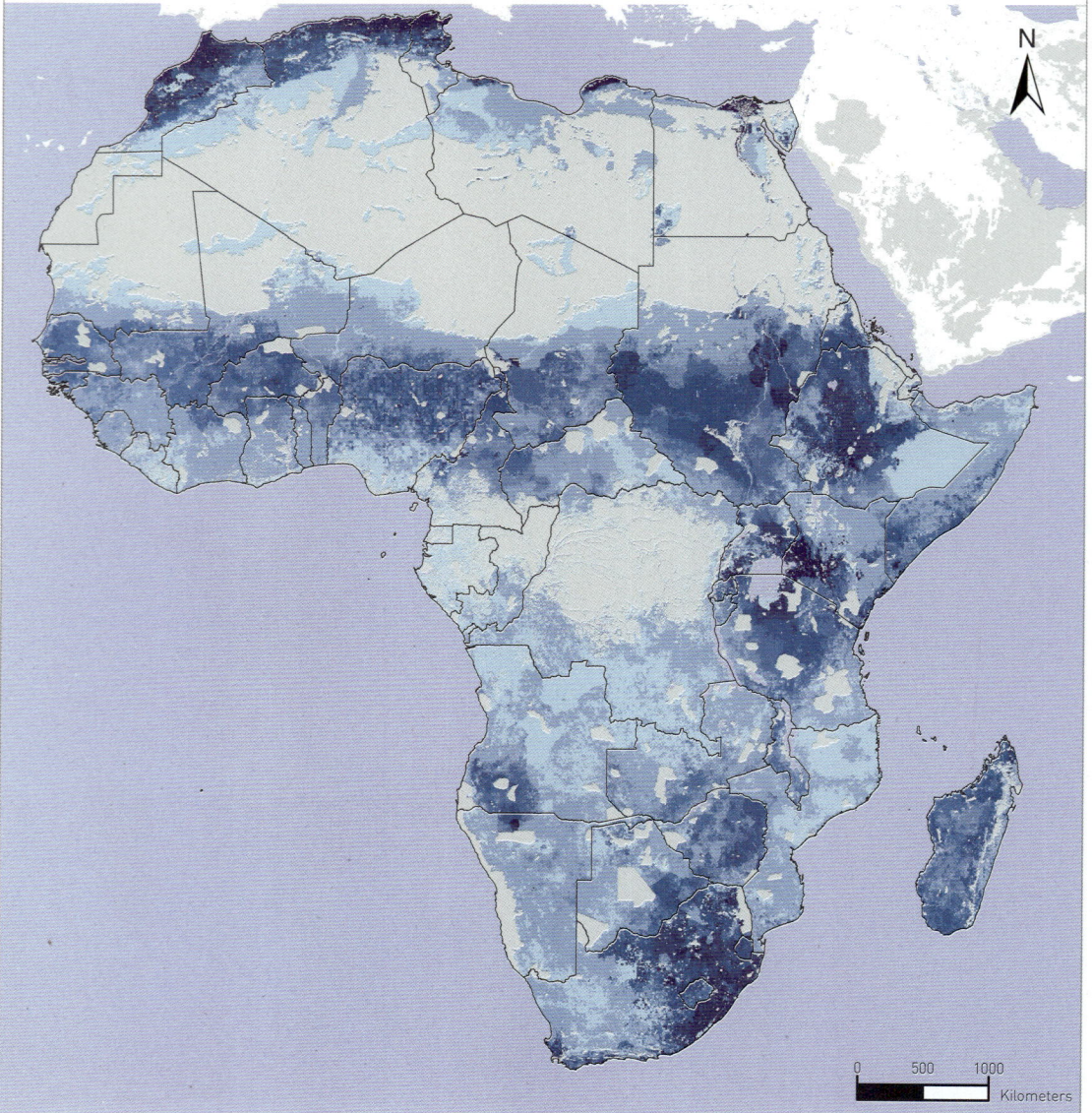

Milk off-take (kg per square km)

←1	500–1 000	5 000–10 000	International boundary
1–250	1 000–2 500	→10 000	Water
250 -500	2 500–5 000		Unsuitable for ruminant

7.12 NATIONAL-LEVEL ESTIMATES OF SHEEP PRODUCTION IN THE MIDDLE EAST

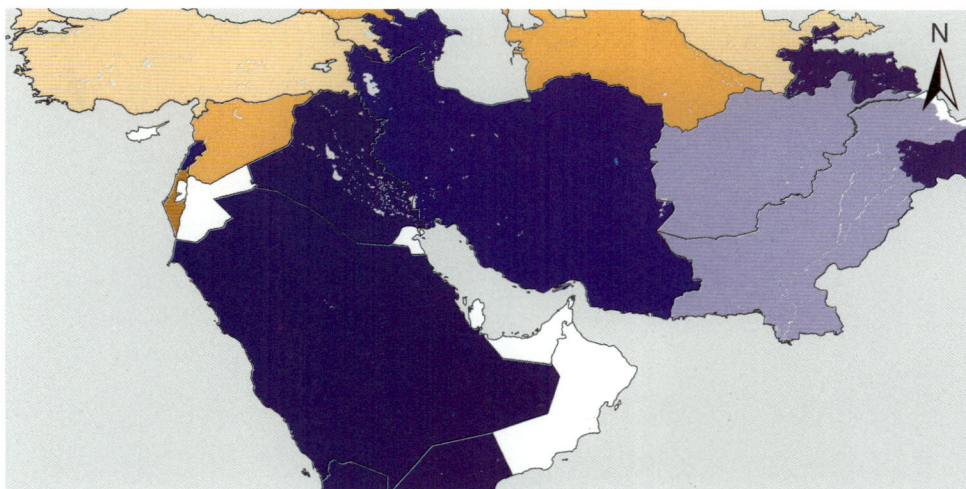

Tonnes per animal (year 2000)

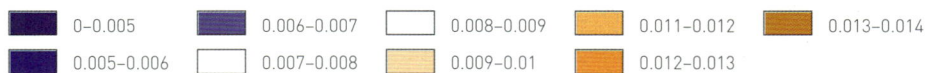

- 0–0.005
- 0.005–0.006
- 0.006–0.007
- 0.007–0.008
- 0.008–0.009
- 0.009–0.01
- 0.011–0.012
- 0.012–0.013
- 0.013–0.014

Source: Derived from Wint and Slingenbergh, 2004.

7.13 NATIONAL-LEVEL ESTIMATES OF CONSUMPTION OF SHEEP MEAT IN THE MIDDLE EAST

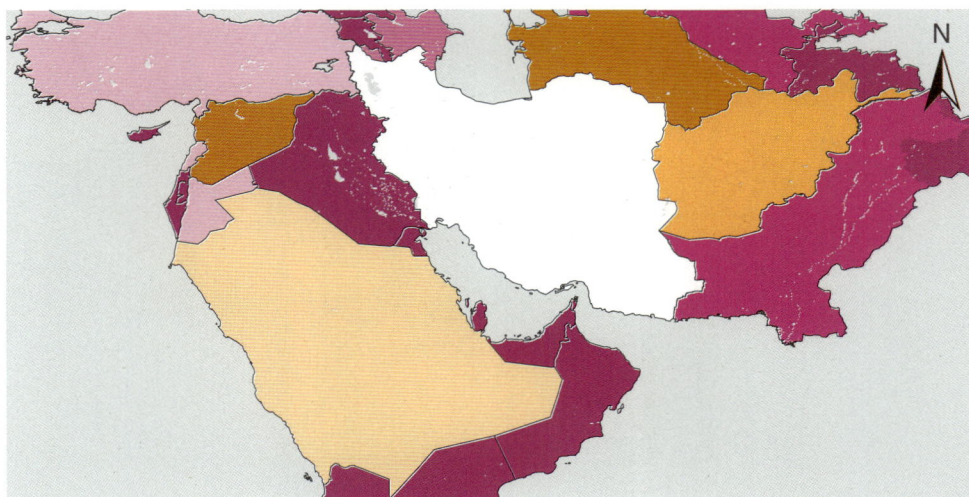

Tonnes per person (year 2000)

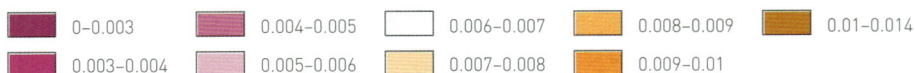

- 0–0.003
- 0.003–0.004
- 0.004–0.005
- 0.005–0.006
- 0.006–0.007
- 0.007–0.008
- 0.008–0.009
- 0.009–0.01
- 0.01–0.014

Source: Derived from Wint and Slingenbergh, 2004.

7.14 MODELLED SHEEP DENSITIES IN THE MIDDLE EAST

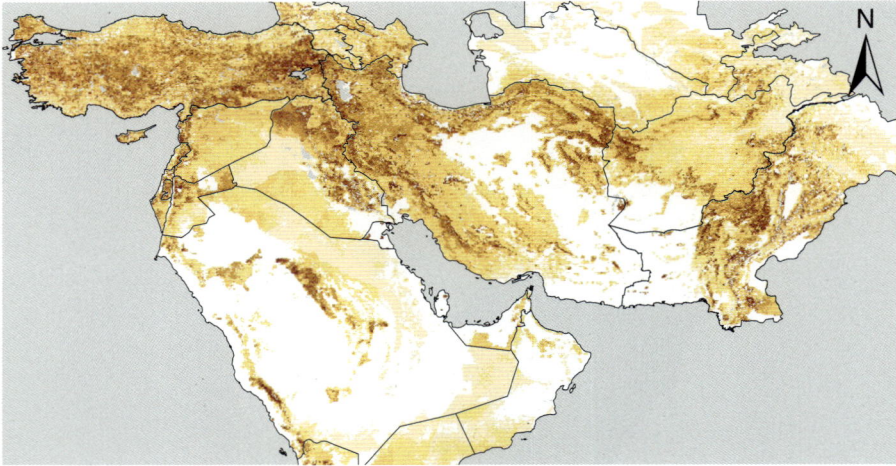

Number per square km

←1	10–20	100–250		International boundary
1–5	20–50	→250		Water
5–10	50–100			Unsuitable for ruminant

7.15 HUMAN POPULATION DISTRIBUTION IN THE MIDDLE EAST

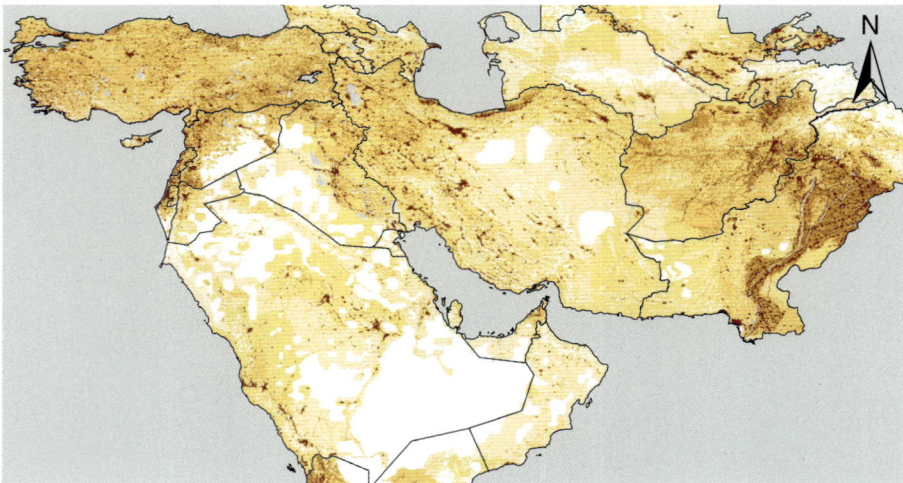

Number per square km

←1	10–20	100–250		International boundary
1–5	20–50	→250		Water
5–10	50–100			Unsuitable for ruminant

Source: Reproduced from LandScan population figures (www.ornl.gov/sci/gist/projects/LandScan).

7.16 PRODUCTION SURPLUS FOR SHEEP MEAT IN THE NEAR EAST (Estimates from the year 2000)

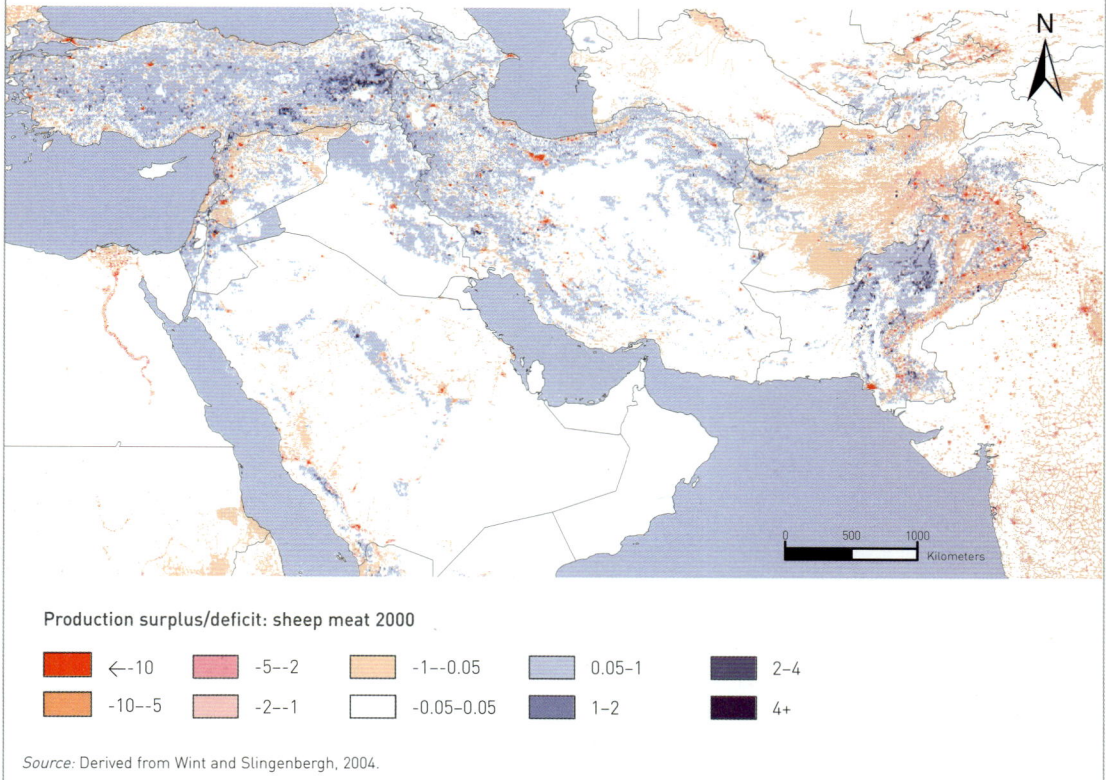

Production surplus/deficit: sheep meat 2000

Colour	Value
(red)	←-10
(orange)	-10--5
(pink)	-5--2
(light pink)	-2--1
(tan)	-1--0.05
(white)	-0.05-0.05
(light blue)	0.05-1
(blue)	1-2
(dark blue)	2-4
(dark purple)	4+

Source: Derived from Wint and Slingenbergh, 2004.

nation has exported livestock products, and from which countries it has imported them. However, available data are not exhaustive and focus more on products than live animals, many of which are moved across borders without official knowledge or permission – so-called 'informal' trade.

A modelling approach to estimating movements of livestock and livestock products is shown here (Wint and Slingenbergh, 2004). This is based on the assumption that national levels of animal production, expressed as kilograms of product per animal and calculated from figures given in FAO, 2003, can be applied to subnational resolution animal distributions to estimate the distribution of animal production per species. An index of the demand for livestock products can be produced by multiplying the calculated country-level demand per person, also provided in FAO, 2003, by subnational

resolution human population data (derived from the LandScan coverages[46]).

These estimates are illustrated in Figures 7.12 to 7.15, which give the components of a production/consumption balance approach to estimating movements of livestock and livestock products in the Near East. Figure 7.12 shows estimates of sheep production per animal (from FAO, 2003) and Figure 7.13 the GLW sheep distribution; combining these gives an estimate of production. Similarly, Figure 7.14 estimates the per capita rate of consumption of sheep meat (from FAO, 2003) and Figure 7.15 shows the LandScan distribution of people; combining these data gives the 'demand' side of the equation. (Obviously, in each case, production and consumption are assumed to be

[46] http://www.ornl.gòv/sci/gist/projects/LandScan

7.17 FARMING SYSTEMS IN SUB-SAHARAN AFRICA MOST LIKELY TO SUPPORT SIGNIFICANT CATTLE MOVEMENTS

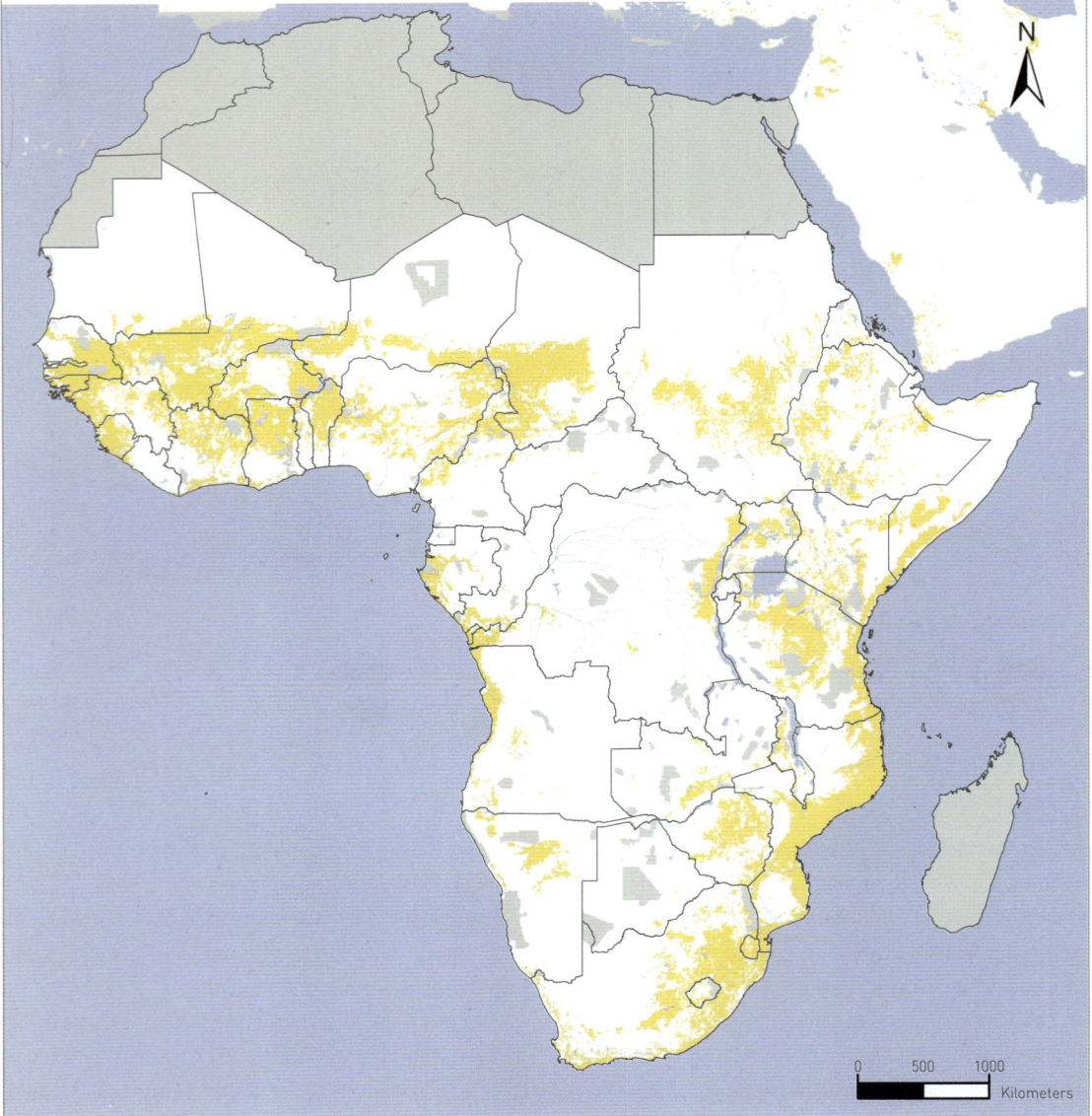

Source: Derived from Wint and Sumption, 2005.
Note: Protected areas and countries outside the analysis are masked in grey.

equal across each country: a limitation caused by the detail of the available figures.)

Combining these production and demand indices, as illustrated in Figure 7.16, produces a production balance map. Not surprisingly, this demonstrates a general surplus in rural areas and a deficit in heavily populated areas in and close to cities. However, it also highlights larger areas where a general surplus is indicated and those where a general deficit is indicated, suggesting movements

7.18 MONTHLY PASTURE AVAILABILITY IN THE NEAR EAST FOR THE YEAR 2002

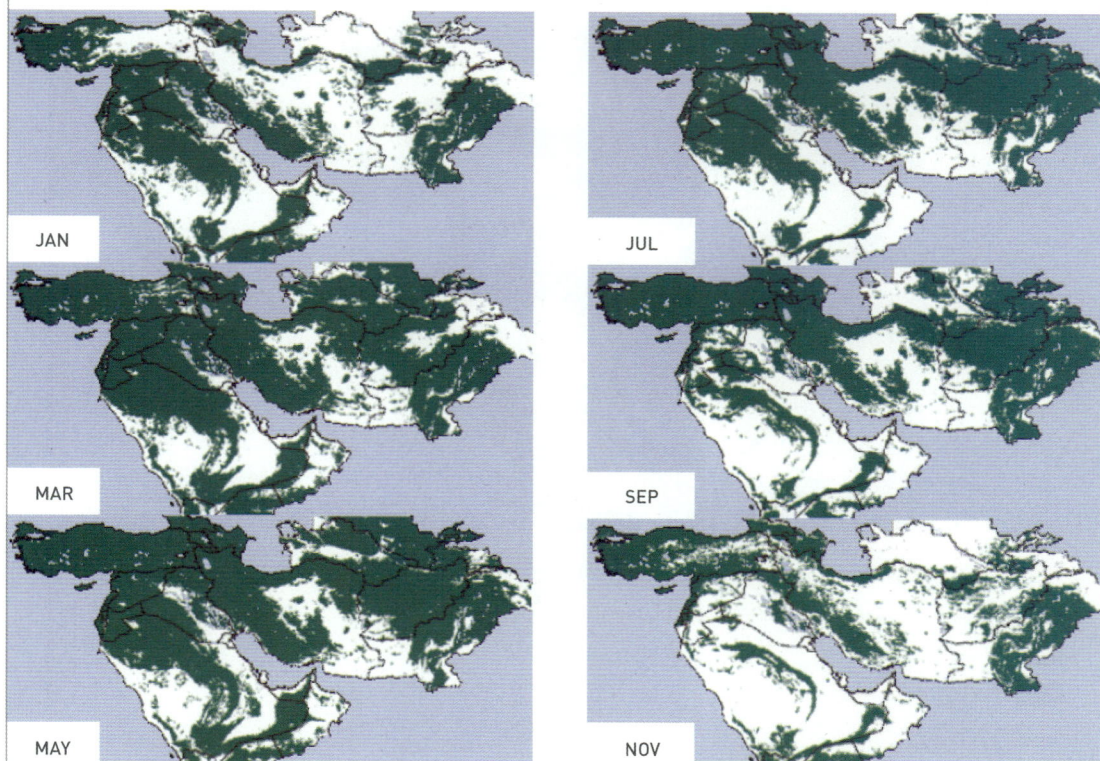

JAN

MAR

MAY

JUL

SEP

NOV

Source: Adapted from Wint, 2003.

of livestock and livestock products from the surplus to the deficit areas. These modelled estimates need to be validated against trade statistics.

LIVESTOCK MOVEMENT

A feature of many forms of animal agriculture is the movement of livestock, either to take advantage of seasonally available resources or for trade, both within and among countries.

As discussed in the previous section, livestock trade statistics are notoriously unreliable in many parts of the world. While some countries have begun to maintain detailed databases of internal livestock movements, such as the Cattle Tracing System in the United Kingdom, they must be seen as the exception. Because a reliable database of global

livestock movements has yet to be established, any tentative assessment must perforce rely on the use of indices, proxies or indicators.

In this context, the production balances described in the previous section may be used as indicators of trade-related movements of livestock or livestock products from areas of production surplus to areas where demand exceeds supply. These trade indicators are most likely to be associated with animal movements where there are substantial areas of demand and production surplus adjacent to one another, or where production surpluses are very high. Other proxies of animal movements may be implied from predominant husbandry systems. For example, areas with significant cattle densities but relatively few people and/or little cultivation,

7.19 SEASONAL PASTURE AVAILABILITY IN THE NEAR EAST FOR THE YEAR 2002

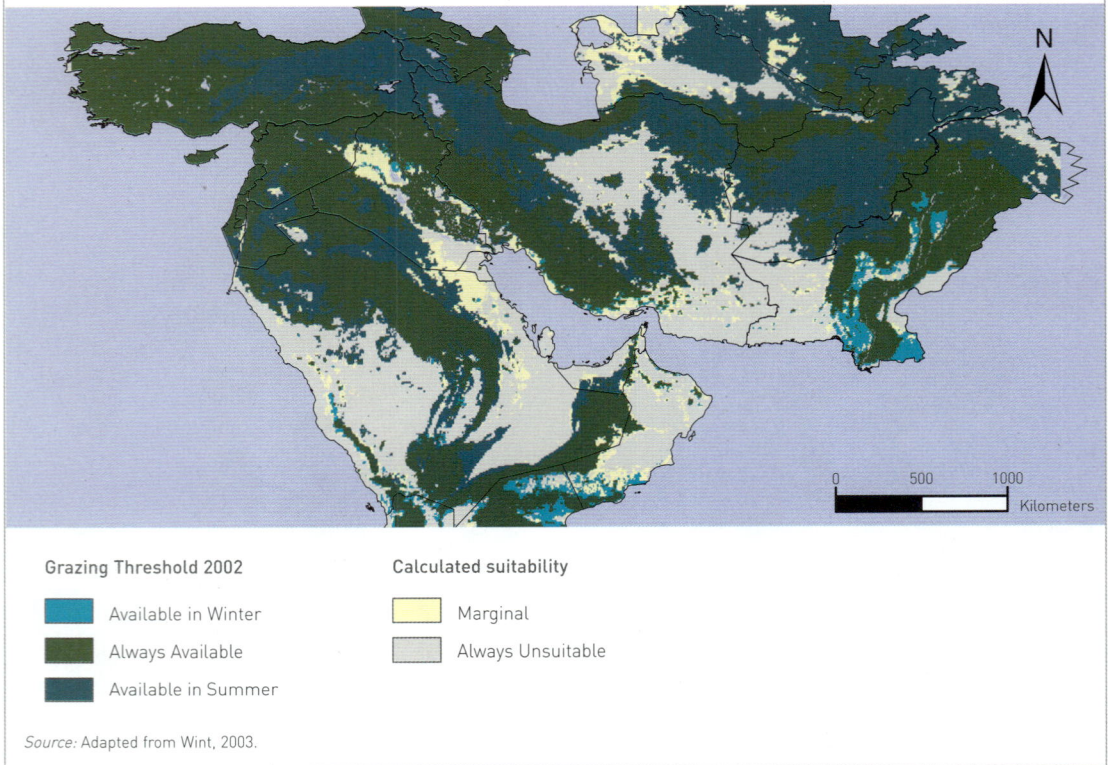

Grazing Threshold 2002

- Available in Winter
- Always Available
- Available in Summer

Calculated suitability

- Marginal
- Always Unsuitable

Source: Adapted from Wint, 2003.

are indicative of transhumant pastoral production. Figure 7.17, for example, shows the distribution of African farming systems (defined in Wint *et al.*, 1999) that satisfy these criteria and are thus likely to support significant cattle movements.

Another way to infer movement, particularly for transhumant animals, is to evaluate the seasonal distribution of pasture for grazing.

Monthly integrated NDVI values are directly related to primary production (Tucker and Sellers, 1986). It therefore follows that an appropriate NDVI threshold could be identified, below which no pasture is available for grazing. Figure 7.18, for example, adapted from Wint, 2003, shows the monthly suitability of pasture for grazing in the Near East, estimated from monthly NDVI data derived from satellite imagery. Areas where the NDVI falls below a specified threshold are masked as unsuitable for grazing during the month in

question. Figure 7.19 summarizes this, showing which areas support grazing only during the winter, which support grazing only during the summer, and those areas where pastures are suitable for grazing throughout the year.

It may be possible also to reinterpret the legends of these maps in terms of seasonal stock movements. Areas where pasture is available only during the winter or summer are less likely to support permanent animal populations. Thus, seasonal movements of animals are likely to occur in the autumn into areas where pasture is available only during the winter, and in the spring into areas where pasture is available only during the summer, with return migrations at the end of those seasons. Animals remaining in marginal areas when there is no pasture available are likely to be fed on stored or imported feed.

7.20 DISTRIBUTION OF TSETSE IN THE HORN OF AFRICA

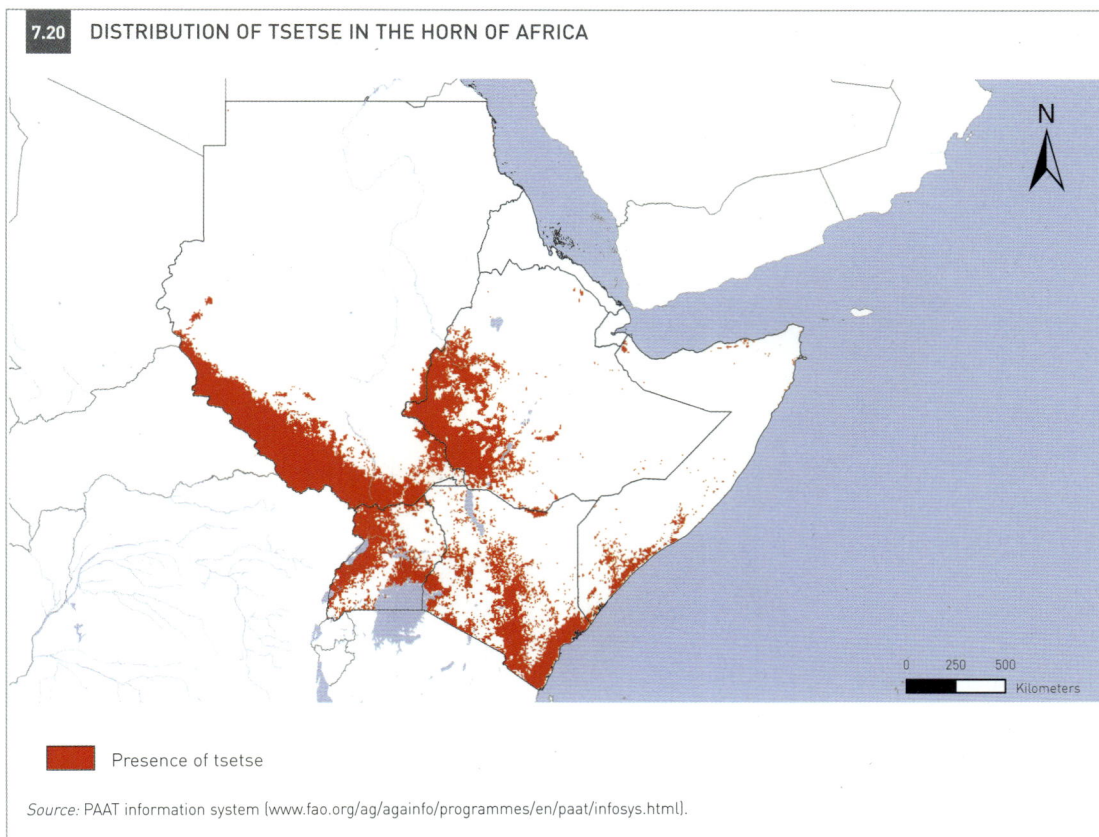

Presence of tsetse

Source: PAAT information system (www.fao.org/ag/againfo/programmes/en/paat/infosys.html).

LIVESTOCK DISEASE ASSESSMENT

Livestock distribution maps are an essential component of any spatial evaluation of the impact of livestock disease, and therefore of livestock disease interventions.

At its very simplest, one can overlay the distribution of a disease or a disease vector on the livestock distribution maps to estimate the numbers of animals that are at risk. For planning applications, such evaluations can be stratified by administrative areas, such as province or district boundaries, for example, to assist with allocations of funds, distribution of vaccines, or placement of particular livestock services. Another useful type of stratification is the production system.

Figure 7.20 provides an estimate of the distribution of tsetse in the Horn of Africa. This was produced by combining the modelled probabilities of presence of the three major groups of tsetse,

available as GIS layers from the Programme Against African Trypanosomiasis (PAAT) information system[47]. These were then combined with the cattle distribution in the Horn of Africa (Figure 7.21 – but adjusted to match FAOSTAT 2005 national totals) and the Thornton *et al.*, 2002, livestock production systems (Figure 7.22) to extract the figures given in Table 7.3.

These figures would suggest that, of the approximately 100 million cattle in the Horn of Africa, some 17 percent are potentially at risk from trypanosomiasis. The greatest absolute numbers of cattle at risk (some 4 million) are in the arid/semi-arid pastoralist areas. However, these represent only 15 percent of total animals in these systems, which cover vast areas of East Africa. Large numbers of cattle, more than 3 million,

[47] http://www.fao.org/ag/againfo/programmes/en/paat/infosys.html

7.21 MODELLED CATTLE DENSITY IN THE HORN OF AFRICA

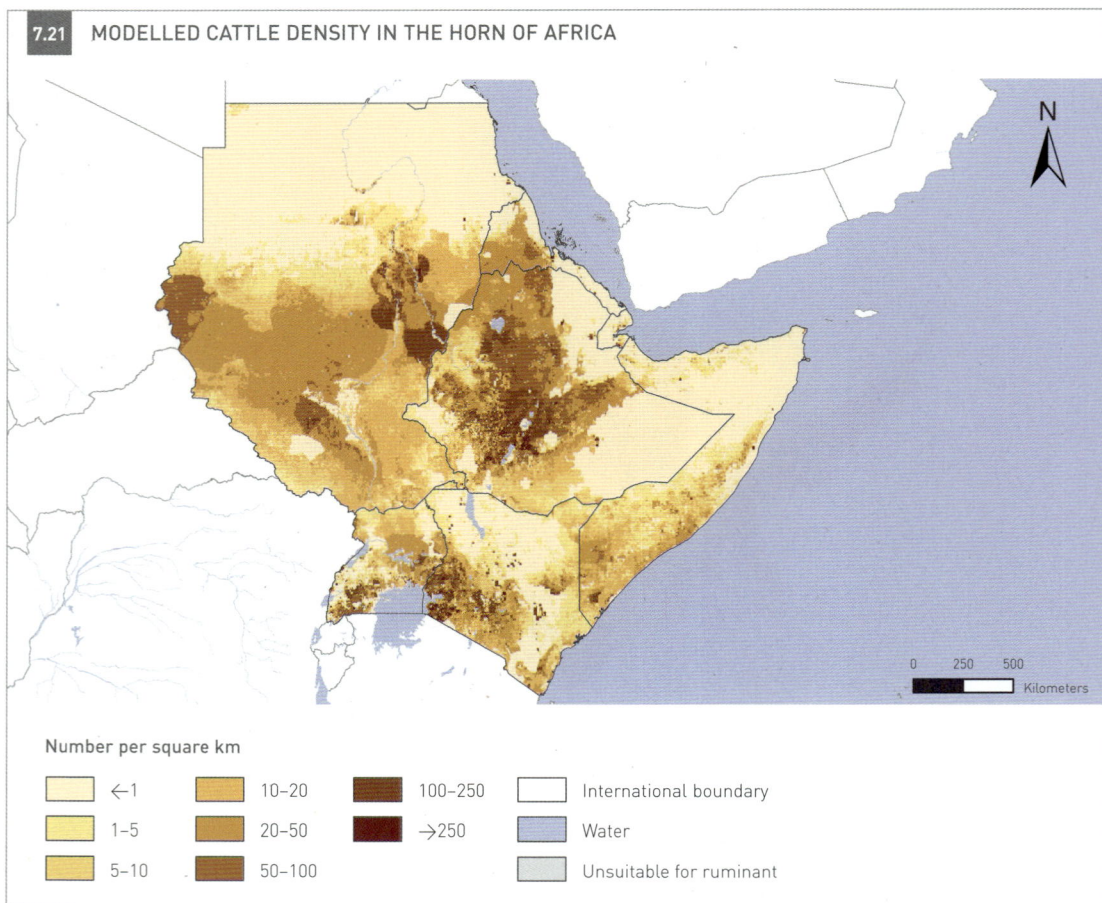

Number per square km

←1	10–20	100–250
1–5	20–50	→250
5–10	50–100	

International boundary
Water
Unsuitable for ruminant

---

are also at risk in the rainfed mixed (arable and livestock) farming areas in the humid/sub-humid zones. In this latter production system, however, they represent three-quarters of the total number of cattle, and therefore interventions might be more appropriately targeted here.

More sophisticated approaches exist to livestock disease-risk or impact assessment. For example, in sub-Saharan Africa, the average sero-prevalence of brucellosis is estimated at 16.2 percent; elimination of the disease would improve fertility rates and calf mortality rates. FAO, 2002a, used the LDPS-2 model described in Section 7.4, stratified by agro-ecological zones, to estimate the potential increase in milk and meat off-take that would result from brucellosis control (on the assumption that cattle population growth rates would remain unchanged).

By using estimates of the value of milk and beef, US$ 0.20/l and US$ 2.00/kg, respectively, they were able to map the estimated financial benefits of brucellosis control. Shaw *et al.*, 2006, have further advanced this approach by incorporating the livestock movement models described in Section 7.2 to map the potential benefits of trypanosomiasis control interventions in West Africa over a 20-year period.

LIVESTOCK DISEASE-RISK MAPPING

Knowledge of livestock distributions is an essential component of livestock disease-risk mapping. Two examples that incorporate livestock densities directly into disease-risk maps are given here: the first relates to bovine tuberculosis (BTB) in the United Kingdom and demonstrates the use of

7.22 LIVESTOCK PRODUCTION SYSTEMS IN THE HORN OF AFRICA

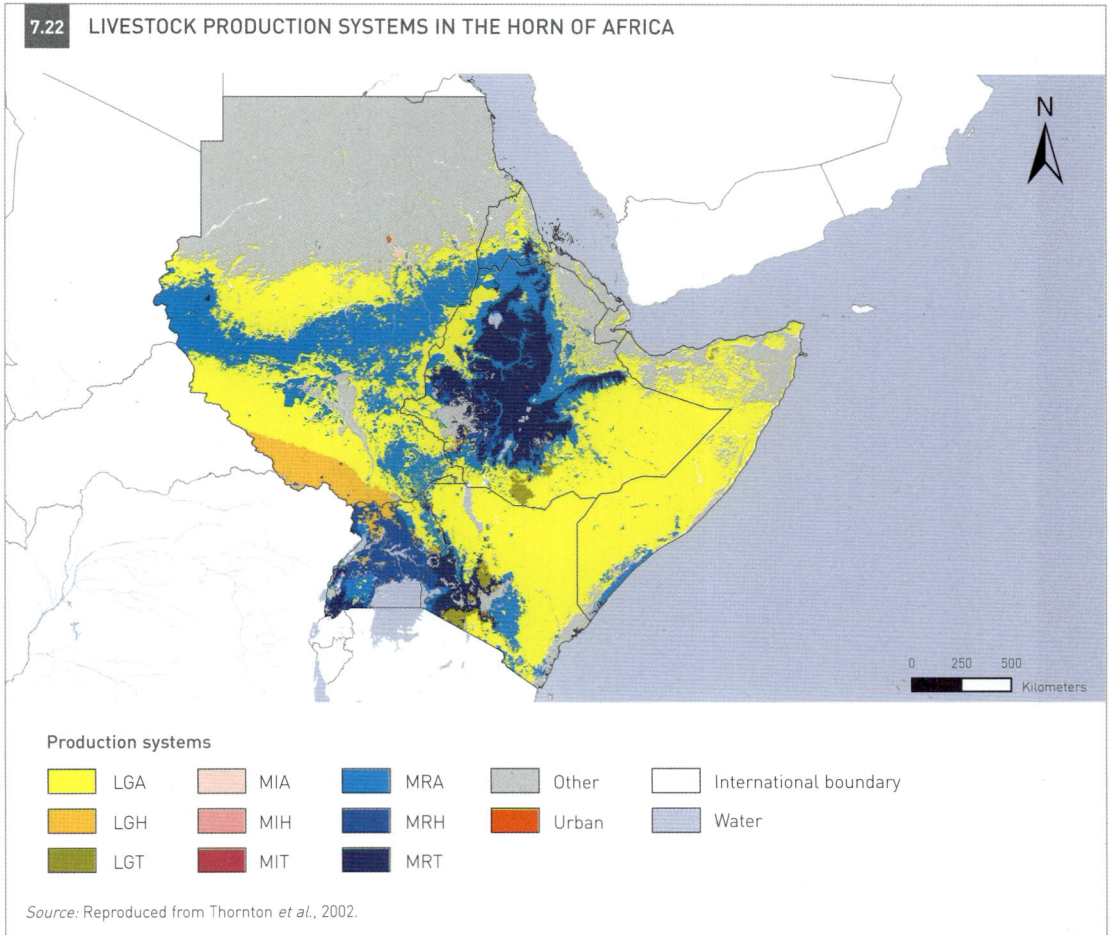

Production systems

LGA	MIA	MRA	Other
LGH	MIH	MRH	Urban
LGT	MIT	MRT	

International boundary

Water

Source: Reproduced from Thornton *et al.*, 2002.

livestock densities in determining disease distribution; the second concerns global risk assessments of FMD. A third example, the global spread of bird 'flu, whilst not using livestock distributions directly to model risk, illustrates how powerful such maps can be in helping to understand disease spread and to present information to the general public in an accessible and appealing way.

BTB in the United Kingdom

Since the mid-1980s, BTB has been spreading in England and Wales and, by 2004, was found throughout southwest and central England, and in eastern and southern Wales. Attempts to model distribution of the disease have highlighted a range of factors that are reliable predictors of the presence or absence of the disease, of which seasonal climatic factors (Wint *et al.*, 2002) and animal movement-related parameters (Gilbert *et al.*, 2005) are the most effective. Animal densities must also be incorporated into the models, both as a mask to delimit areas where the disease may occur (and thus where it is necessary to monitor it), but also as one of the main determinants of disease presence within its range. Figure 7.23 for example, shows the predicted distribution of BTB in the United Kingdom for 2003, with the actual distribution inset, adapted from Wint *et al.*, 2002.

FMD status

Animal density distributions can be used more directly to assess disease risk in areas where

TABLE 7.3 NUMBERS OF CATTLE WITHIN THE DISTRIBUTION OF TSETSE AND THEREFORE POTENTIALLY AT RISK FROM TRYPANOSOMIASIS IN THE HORN OF AFRICA, BY COUNTRY AND LIVESTOCK PRODUCTION SYSTEM, ADJUSTED TO FAOSTAT 2005 NATIONAL TOTALS

Production System	Djibouti	Ethiopia	Eritrea	Kenya	Somalia	Sudan	Uganda	System Total
LGA	0	425 545	0	865 345	344 500	2 694 820	39 280	4 369 490
		11%		30%	8%	16%	15%	15%
LGH	n.a.	42 770	n.a.	n.a.	n.a.	958 060	75 300	1 076 130
		98%				99%	36%	88%
LGT	n.a.	16 190	0	130 830	0	6 210	4 700	157 930
		7%		25%		98%	44%	20%
MIA	n.a.	0	0	n.a.	53 030	0	n.a.	53 030
					40%			11%
MRA	0	1 277 160	0	1 582 800	146 440	622 200	449 810	4 078 410
		15%		77%	39%	4%	29%	13%
MRH	n.a.	1 280 250	n.a.	889 965	n.a.	8 230	1 635 930	3 814 375
		100%		88%		100%	59%	75%
MRT	n.a.	2 209 075	0	784 360	n.a.	18 930	251 650	3 264 015
		10%		20%		95%	30%	12%
Other	0	327 430	0	531 800	128 230	93 910	187 230	1 268 600
		26%		36%	20%	3%	43%	18%
Country Total	0	5 578 420	0	4 785 110	672 200	4 402 360	2 643 900	18 081 990
		15%		40 %	13%	11%	43%	17%

Notes: Livestock production system data were taken from Thornton *et al.*, 2002. The percentages indicate the number of animals at risk as a proportion of the total number of cattle occurring in that stratum (from Table 7.2).
'n.a.' indicates that system does not occur in a country.

7.23 BTB RISK IN THE UNITED KINGDOM, 2003

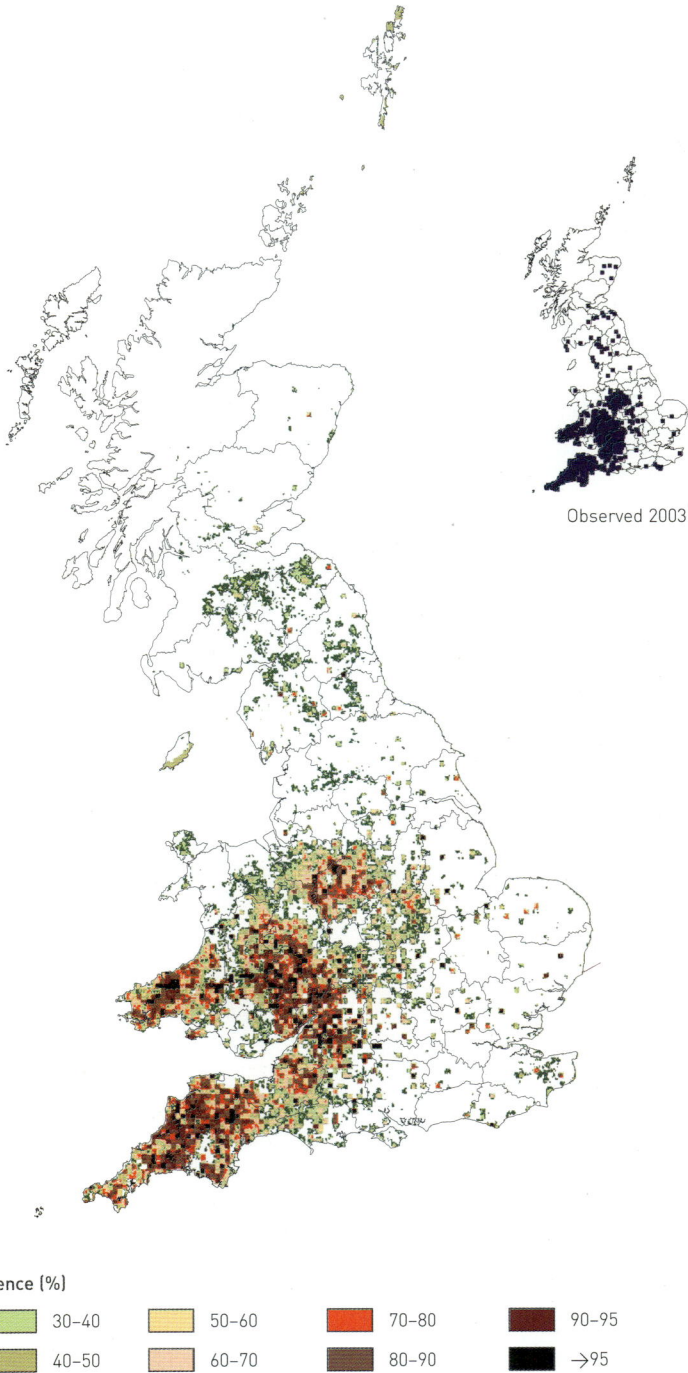

Observed 2003

Probability of presence (%)

←20	30–40	50–60	70–80	90–95
20–30	40–50	60–70	80–90	→95

Source: Adapted from Wint *et al.*, 2002.

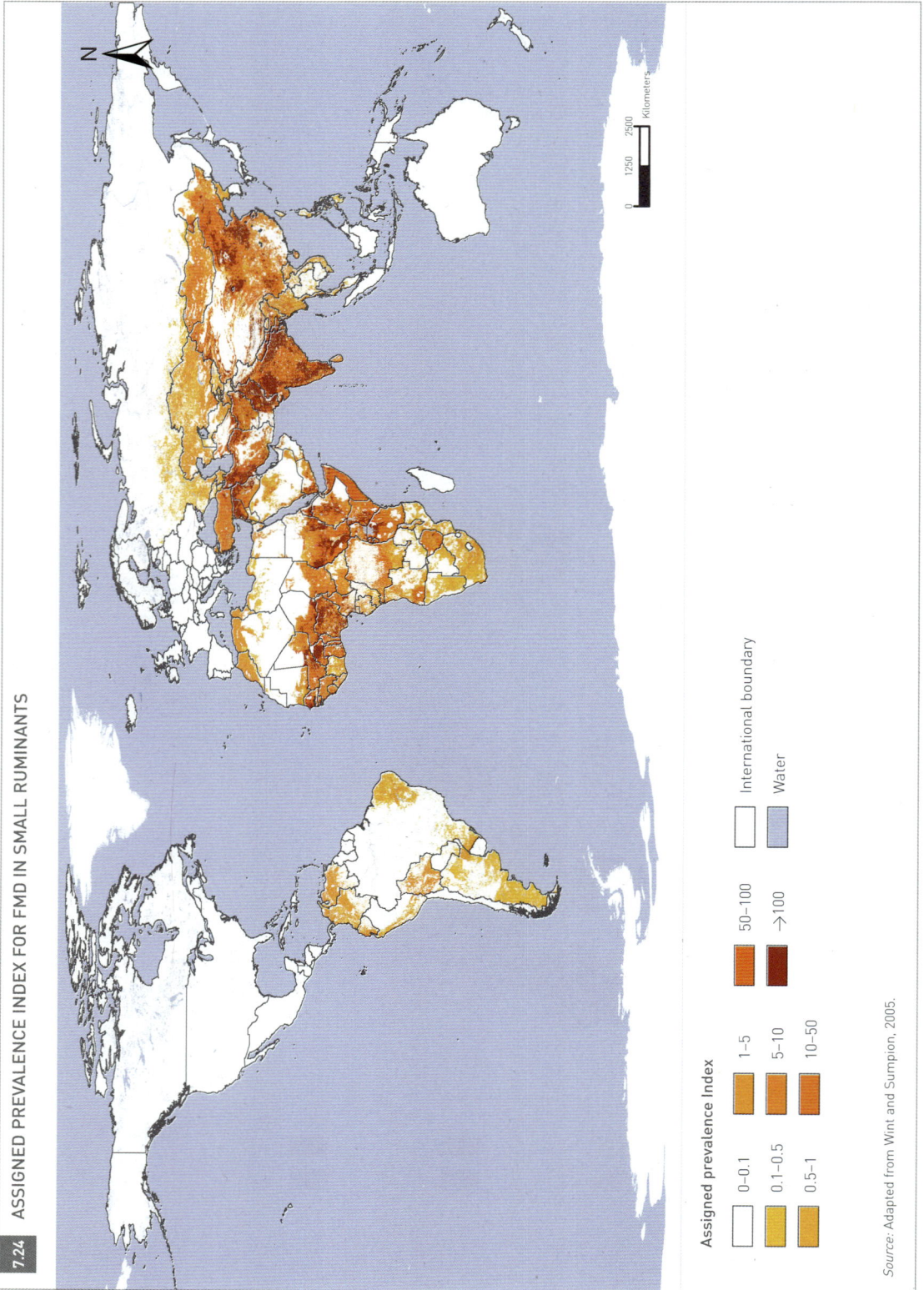

7.24 ASSIGNED PREVALENCE INDEX FOR FMD IN SMALL RUMINANTS

Assigned prevalence Index

0–0.1	1–5	50–100
0.1–0.5	5–10	→100
0.5–1	10–50	

International boundary

Water

Source: Adapted from Wint and Sumpion, 2005.

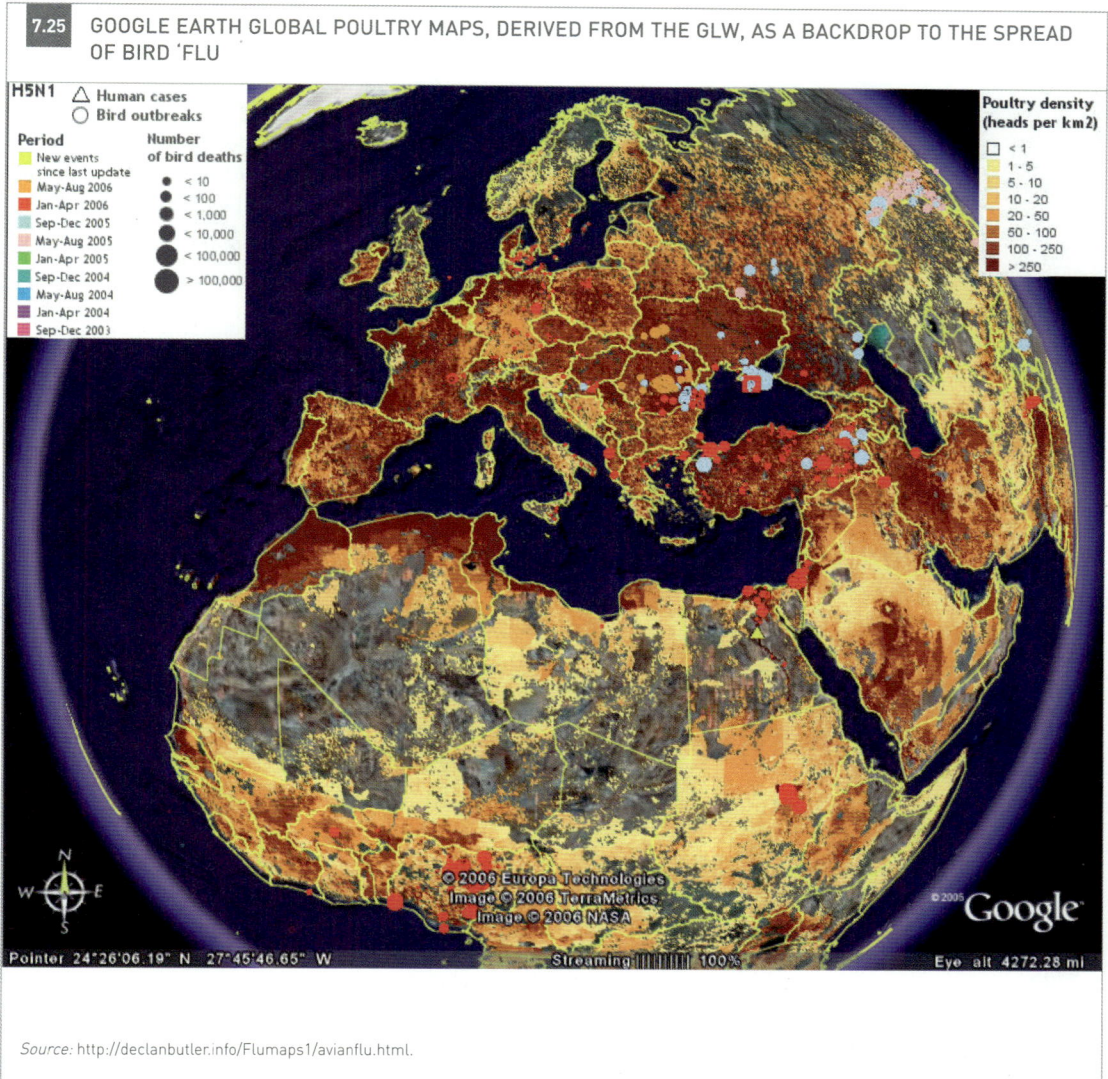

7.25 GOOGLE EARTH GLOBAL POULTRY MAPS, DERIVED FROM THE GLW, AS A BACKDROP TO THE SPREAD OF BIRD 'FLU

Source: http://declanbutler.info/Flumaps1/avianflu.html.

direct monitoring is unreliable or sparse. Reliable surveillance data for FMD, on which to base prevalence estimates, is only available for a small proportion of countries. New approaches are therefore required to estimate the potential disease burden in countries with large animal populations that may hold a significant proportion of the global pool of FMD virus.

In an attempt to overcome the lack of quantitative information, FMD surveillance data for 'representative' country or husbandry systems has been used to generate annualized incidence values

that may then be applied to countries with the same, or similar, conjectural FMD status (Wint and Sumption, 2005).

A constant incidence was then applied for all countries within the same zone of conjectural FMD status. This assigned incidence index was combined with the density distributions of each species to derive an indicative prevalence index within countries. The resultant global FMD prevalence index for cattle is shown in Figure 7.24.

In such an approach, the main variable driving the number of cases is the population at risk. Thus the

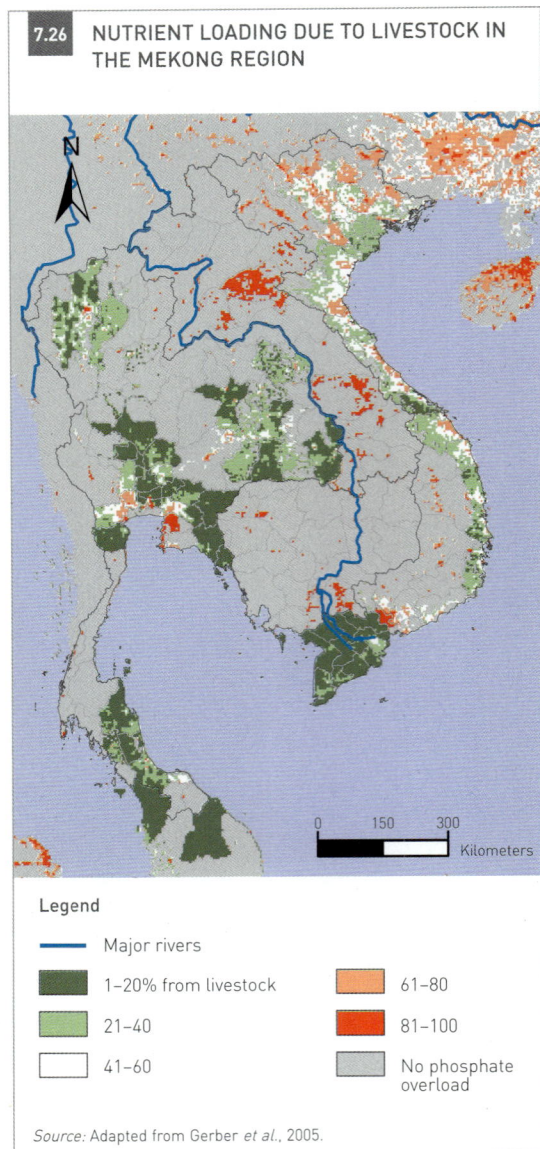

7.26 NUTRIENT LOADING DUE TO LIVESTOCK IN THE MEKONG REGION

Legend

— Major rivers

■ 1–20% from livestock

■ 21–40

□ 41–60

■ 61–80

■ 81–100

■ No phosphate overload

0   150   300
Kilometers

Source: Adapted from Gerber *et al.*, 2005.

The global spread of bird 'flu

Animal density maps may be also used as an aid in interpreting disease distributions. Many epidemiologists, governments and the public in general closely followed the spread of bird 'flu from Southeast Asia during 2005 and 2006. A very widely available example is the use of the GLW poultry layers as a backdrop to the maps of disease outbreaks in poultry, made available via Google Earth by Declan Butler[48] (shown in Figure 7.25). This clearly illustrates the coincidence of poultry outbreaks in the Near East, West Asia and Africa with high poultry densities.

ENVIRONMENTAL IMPACT ANALYSIS

Livestock affect the environments they inhabit in a variety of different ways, for example, through overgrazing and soil erosion; production of methane and other greenhouse gases; nutrient recycling in extensive systems; excessive nutrient concentrations in the effluent from intensive systems; influence on land use; and displacements of wildlife. Efforts to quantify and monitor these effects can only be successful if estimates of livestock numbers are both reliable and available at appropriate resolutions.

The FAO Livestock, Environment and Development initiative[49] explores many aspects of the impact that livestock have on the environment (FAO, 2006a). For example, recently published studies (Gerber *et al.*, 2005) have provided nutrient balance maps of the Mekong region of Southeast Asia, using phosphate as an indicator. Estimated excretion values per animal for each species were applied to the livestock-density distribution models to provide an index of livestock-generated phosphate per square kilometre. A similar procedure was used to estimate phosphate uptake by crops. An input value was calculated for the rate of fertilizer application, by apportioning national fertilizer use only to regions supporting high-yielding crops, as indicated by subnational cropping data. The three

significance of countries will depend on the relative size of their livestock populations. Although this may under-estimate particularly high- or low-risk zones, it avoids the under-representation of some endemic countries with large livestock populations. Focusing on the application of annualized incidence rates may mask particular risks from antigenic divergence – the risk of 'exotic FMD types' – but more systematic study may reveal patterns of emergence that would refine risk assessment.

[48] http://declanbutler.info/flumaps1/avianflu.html
[49] http://www.lead.virtualcentre.org

TABLE 7.4 CRITERIA WEIGHTS FOR SELECTING PRIORITY AREAS FOR TRYPANOSOMIASIS CONTROL IN UGANDA

Factors	Weights
Density of the poor livestock keepers	0.2562
Trypanosomiasis risk index	0.5030
Length of growing period	0.0559
Cattle density	0.1546
Percentage crop cover	0.0304

Source: FAO, in preparation.

7.27 TRYPANOSOMIASIS CONTROL PRIORITY MAP FOR UGANDA

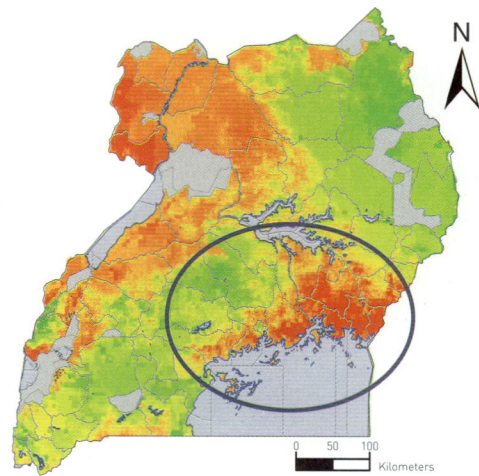

Legend

- National Parks
- Administrative boundaries
- Water bodies

Priority
- High
- Low

Source: FAO, in preparation.

components were then combined to estimate the phosphate balance. Areas of nutrient overloading were identified using the phosphate balance model. Then, by combining this with gridded livestock distribution data, it was possible to assess the relative contribution to nutrient loading made by livestock species, as illustrated in Figure 7.26.

SPATIAL TARGETING OF INTERVENTIONS

Many of the examples given above will be implemented to aid decision-making, for example, by field operatives and policy-makers. This will be done mostly with some component of spatial targeting, for example, where disease interventions might be best placed, or where it might be most important to mitigate environmental impacts. In this final example, it is demonstrated how such spatial targeting can be taken a step further by incorporating livestock data with other relevant spatial information in a decision-support model.

In planning trypanosomiasis control, two overriding questions are essentially involved: how to control, and where to prioritize efforts. The answer to the first depends on a multitude factors, such as the relative costs of different interventions, whether the objective is to control the disease or to eliminate the vector, and the local disease epidemiology and biology. To answer the second question requires a very clear objective, and the decision criteria will be often determined by economic rather than technical considerations. In Uganda, as in a number of other African countries, renewed efforts are being made

to control trypanosomiasis, influenced largely by the Pan African Tsetse and Trypanosomiasis Eradication Campaign (PATTEC), which advocates wide-scale eradication of tsetse. Whichever methodology, or combination of technologies, is ultimately used to intervene there is a clear need to target interventions appropriately.

In a collaborative project between the Coordinating Office for the Control of Trypanosomiasis in Uganda, FAO and ILRI, a multicriteria evaluation technique – weighted linear combination – was used to combine relevant spatial data to identify priority areas 'to control animal trypanosomiasis for the alleviation of poverty' (FAO, in preparation; and see Robinson *et al.*, 2002, for a detailed description of the methodology). Five relevant criteria were created and digitized to produce standardized maps for GIS analysis: (i) density of poor livestock keepers; (ii) trypanosomiasis risk index; (iii) LGP;

(iv) cattle density; and (v) percentage crop cover (details on how these were produced are given in FAO, in preparation). Weights were assigned to these criteria by decision-makers and other stakeholders in the livestock sector in Uganda, through an iterative process of workshops. A set of consensus weightings was eventually reached, as shown in Table 7.4.

The priority map given in Figure 7.27 was produced by summing the weighted input criteria.

A precursor to the map in Figure 7.27 was used to help target PATTEC interventions in Uganda: the red areas (high priority) contained within the blue ellipse in Figure 7.27 were selected as the zone where the initial activities would be implemented. Further GIS analysis reveals that this area contains some 754 000 head of cattle and a rural human population of some 5 million, of which about 2.6 million live below the US$ 1 per day threshold. Hence one can start to use these data to estimate the types and magnitude of impacts that might be achieved by targeted interventions.

8 Challenges and future directions

In its many forms, livestock production is an important component of most agricultural economies. Yet the livestock sector is frequently marginalized in terms of development priorities and allocation of resources, despite the ever-present media headlines highlighting the possible dangers of mad cows, FMD, bird 'flu and other emerging zoonotic diseases. The sector, particularly in low-income countries, is frequently perceived as intractable, dispersed and often located in the most remote rural areas away from towns and administrative centres – and therefore difficult to enumerate, monitor and develop effectively.

The production of digital livestock maps has opened up exciting possibilities, and will allow for a number of types of analyses that, until now, have been difficult if not impossible to carry out. Whilst such maps are a significant step forward in making global livestock statistics available, there are several priorities for further investigation and development.

It is quite clear from a detailed inspection of the metadata accompanying the GLIS that there is enormous variability in spatial resolution and species definition (particularly of poultry), and a wide range in the date of origin of the input data. It is to be hoped that increasing automation of the data management and modelling process will facilitate much more frequent updates, so that in due course new census or survey data can be incorporated into the Oracle database as the information becomes available and all products updated automatically. In reality there are likely to be delays as a result of validation and data cleaning procedures that tend to be specific to each data source. However, as subnational data reporting becomes more common, so the data formats should become more standardized

It is also quite clear that the modelling approaches used here are better suited to some species and production systems than to others. The origins of this environmental approach to livestock distribution modelling are to be found with cattle populations in Africa, where production is closely coupled with the land and such approaches are highly appropriate. In marked contrast, for example, are the often land-detached intensive poultry production systems of Europe and North America, for which environmental approaches are likely to be less well suited.

One province of Germany, for example, contains 3 million chickens; half a million are distributed among 'smallholders', for which this approach to distribution mapping is appropriate, but the remaining 2.5 million are held on only two farms. While a comprehensive map may try to integrate these two types of distribution by using environmental modelling overlain with 'raw' high-resolution data for the intensively reared populations, it may prove more appropriate to map populations from the two systems separately. The problem, however, lies in the level of detail in the reported data, which often precludes making such important distinctions.

Access to better data may mean that other methods, such as the United States Department of Agriculture's Farm Animal Demographics Simulator (FADS) (Freier *et al.*, 2007), could be used to disaggregate livestock in intensive holdings. The basic concept of FADS is to take an area, such as an administrative boundary, for which livestock population data are available and then to remove spatially all areas where the commodity in question would not be found (e.g. lakes, rivers, wetlands, parks, nature reserves, military land, and so on) – in much the same way that unsuitable areas are masked in the GLW process. Farms are then distributed in the remaining area, based on a series of weighting factors that are known to influence their location, such as road access (weighted in terms of suitability for large vehicles delivering feed or transporting live animals) and distance from cities (i.e. markets). Other factors that may relate to the location of farms producing particular commodities are included in the model as they become available.

While Table 6.1 shows that there is already good coverage of the major species and species groups, this could still be improved upon. Broader species coverage would include, for example, yaks, camelids and equines, and use a more consistent definition of the various poultry species. Similarly it would be desirable to have a much better definition of farm types (for reasons discussed above), distinguishing at least between smallholder and industrial production. These distinctions are beginning to be made, inasmuch as a number of countries already provide separate estimates of traditionally managed and 'other' livestock.

From a purely technical perspective, priority must be given to migrating towards a consistent 1 km resolution global (rather than continental) product. That is not to say that in all cases this increased spatial resolution would be reflected in the accuracy of the predictions, where mostly the limiting factor is the quality of the input data and the validity of the statistical model, but there are a number of technical reasons why this is important. First, it would allow us to take advantage of new high-resolution global datasets that may be used for suitability masking and distribution modelling (see, for example, Hay *et al.*, 2006). Second, it would improve the accuracy of the land area estimates, and result in closer correspondence between the raster data and the vector administrative boundary data – all resulting in more accurate livestock density estimates. Third, a single global product would facilitate analysis in relation to the rising number of other standard global 1 km products, and also facilitate automation of the modelling process and therefore the frequency of updates.

Perhaps the most pervasive challenge is to change the target resolution of annual international reporting requirements of agriculture ministries and statistical departments from national to subnational, allowing statistical modelling techniques to be regularly applied to update livestock distribution maps. Such information is usually available from agricultural censuses, and would require a fairly modest investment of resources into the acquisition, collation and analysis of existing data (such as those 'buried' in hard-copy census reports). Many countries now produce detailed digitized subnational agricultural census data (e.g. Brazil, Mexico, the United States) that would require only minimal processing in order to incorporate them into a global subnational resolution archive.

The reporting of subnational data to international agencies by networks would pay immediate dividends, provided, of course, such information was reliably geo-referenced. Some attention should also be given to assessing less credible statistics from regions for which conventional census techniques are inappropriate, and, perhaps, through a limited and carefully targeted monitoring and validation programme. FAO's CountrySTAT[50] is a pilot project launched by FAOSTAT to provide countries with methodologies for compiling, verifying, validating, organizing, analysing and disseminating subnational data related to agriculture and food for the purpose of facilitating data use by national policy-makers and researchers. It is well placed to be the start of such a system, and the data collected could be greatly enhanced by the modelling techniques described here.

Such an initiative, with regular updating and the inclusion of reliability statistics, would provide better livestock-related information for inclusion in poverty, food security and environmental assessments of the type described here. It would also increase the reliability of disease-risk mapping and benefit-cost analysis of disease-control measures.

To date, most attention and effort in livestock mapping has focused on animal numbers or densities, yet their economic importance relates mainly to the value of their products and services. Tentative first steps have been taken in mapping production of cattle meat and milk, but these efforts

[50] http://www.fao.org/es/ess/countrystat/

need to be improved upon and extended to include other species and products and to account for spatial mosaics of livestock production systems.

Livestock must eat to survive and in doing so often eat fodder grown on land that could equally well produce crops that people could eat. There is a potential conflict between land-use for producing animal feed versus its use directly for human food production, especially in ecologically marginal areas with large human and/or livestock populations, but where access to imported feed supplements is limited. Such areas can only be quantified and located if reliable information about livestock numbers, cropped areas and human population density is available at high spatial resolution.

Interactions between animal husbandry and other aspects of agricultural production and renewable natural resources utilization are intimately bound up with and, to a large extent, driven by, economic and social factors that have been largely ignored or avoided by quantitative livestock geography. Until these key elements of animal agriculture can be integrated effectively into a single quantified and geo-referenced framework, monitoring and evaluation of the sector will remain problematic.

As global spatial datasets are now more widely available and diversified, including information on topography, climate, vegetation, land-use, people and livestock, the characterization and mapping of agricultural production systems have become an expanding area of study. There is a real danger, however, that definitions will proliferate and cause confusion rather than clarity, unless a coordinated approach is adopted and objectives are clarified. The increasing availability of quantitative information means that the definition of farming systems can be driven by both quantitative and qualitative data, thereby moving towards higher-resolution mapping rather than the production of homogenous polygons that obscure local heterogeneity.

Much of the preceding discussion has (intentionally) sidestepped the fact that, in many parts of the world, the livestock sector is in a state of flux. Industrial production of pigs and poultry is increas-ing rapidly in the developing world, de-coupling the traditional association between land resources and livestock numbers because large-scale production units are often sited more for efficient access to inputs and transport of products than in terms of land suitability and availability of local natural resources. In contrast, the demand for meat from extensively reared stock is increasingly rapidly in much of the industrialized world.

Furthermore, while human populations and demand for livestock products increase relentlessly, with the accompanying urbanization of human populations and intensification of production, climate change may be about to reshape the agricultural (crops and livestock) geography of the planet. To these trends must be added the inevitable effects of globalization on the movements of animals and foodstuffs, and the spread of existing livestock diseases and the emergence of new ones.

Locating and mapping these trends is crucial to providing adequate decision support for strategic planning, but little has been achieved in these areas to date. The mapping of landless livestock and intensive production units, primarily for pigs and poultry, is crucial to quantifying such trends and, as discussed above, has yet to be adequately addressed, at least at the regional and global levels. This topic's importance is highlighted by the recent realization that the coexistence of traditional and intensive modes of production in urban and peri-urban areas is an increasing cause for concern, as the epidemiological significance of this proximity becomes more apparent and critical to the emergence and spread of zoonotic diseases.

The establishment of the GLIS at FAO represents an important advance in the automation of livestock data acquisition, distribution modelling and dissemination. However, substantially more will need to be done to encourage feedback from national data providers, not only to assess the validity of the outputs but also to return the data with some value added and promote wide interest and use of the global resources to which they have contributed.

9 References

Anderson, J.R., Hardy, E.E., Roach, J.T. & Witmer, R.E. (1976). *A land use and land cover classification system for use with remote sensor data*. 28 pp. Reston, Virginia: US Geological Survey. Professional Paper 964. 28 pp.

Bourn, D. & Wint, W. (1994). Livestock, land use and agricultural intensification in sub-Saharan Africa. Pastoral Development Network 37(a): 1-22. London: Overseas Development Institute.

Bourn, D., Maitima, J., Motsamai, B., Blake, R., Nicholson, C. & Sundstol, F. (2005). Livestock and the environment. In: Owen E. A., Kitalyi, N., Jayasuriya T. Smith (eds): *Livestock and wealth creation: improving the husbandry of animals kept by resource-poor people in developing countries*. Nottingham: Nottingham University Press.

Bourn, D., Wint, W., Blench, R. & Woolley, E. (1994). Nigerian livestock resources survey. *World Animal Review* 78, 49-58. Rome: Food and Agriculture Organization of the United Nations (FAO).

Clarke, R. (ed.) (1986). *The handbook of ecological monitoring*. Oxford: Clarendon Press.

Deichmann, U. (1996). *Asia Population Database 1996*. National Center for Geographic Information and Analysis (NCGIA), University of California, Santa Barbara, as a cooperative activity between NCGIA, Consultative Group on International Agricultural Research and United Nations Environment Programme/Global Resource Information Database (UNEP-GRID). Sioux Falls.

Delgado, C., Rosegrant, M., Steinfeld, H., Ehui, S. & Courbois, C. (1999). Livestock to 2020: The Next Food Revolution. Food Agriculture and the Environment. Discussion Paper 28. Washington: International Food Policy Research Institute, Food and Agriculture Organization of the United Nations and the International Livestock Research Institute.

Dixon, J., Gulliver, A. & Gibbon, D. (2001). Global farming systems study: Challenges and priorities to 2030. Rome and Washington: Food and Agriculture Organization of the United Nations and the World Bank.

Döll, P. & Siebert, S. (2000). A digital global map of irrigated areas. *ICID Journal*, 49, 55-66.

FAO. (1995a). *Programme for the World Census of Agriculture 2000*. Rome: Food and Agriculture Organization of the United Nations.

FAO. (1995b). *Conducting agricultural censuses and surveys*. FAO Statistical Development Series No. 6. Rome: Food and Agriculture Organization of the United Nations.

FAO. (1996). *World livestock production systems: current status, issues and trends*, by C. Seré & H. Steinfeld. Animal Production and Health Paper 127. Rome: Food and Agriculture Organization of the United Nations.

FAO. (1997). *LDPS-2 user's guide*, by L.G. Lalonde & T. Sukigara. Rome: Food and Agriculture Organization of the United Nations.

FAO. (2002a). *Bovine brucellosis in sub-Saharan Africa: estimation of sero-prevalence and impact on meat and milk off-take potential*, by M.J. Mangen, J. Otte, D. Pfeiffer & P. Chilonda. Rome: Food and Agriculture Organization Livestock Policy Discussion Paper No. 8.

FAO. (2002b). *Cattle and small ruminant production systems in sub-Saharan Africa: A systematic review*, by J. Otte & P. Chilonda. Rome: Food and Agriculture Organization of the United Nations.

FAO. (2003). *World agriculture: towards 2015/2030, An FAO Perspective*, by J. Bruinsma. Rome and London: Food and Agriculture Organization of the United Nations and Earthscan.

FAO. (2006a). *Livestock's long shadow: environmental issues and options*, by By H. Steinfeld, P. Gerber, T. Wassenaar, V. Castel, M. Rosales, C. de Haan. Rome: FAO. 390 pp.

FAO. (2006b). *Poverty Mapping in Uganda: An analysis using remotely sensed and other environmental data*, by D.J.R. Rogers, T. Emwanu, T. & T.P. Robinson. Pro-Poor Livestock Policy Initiative Working Paper 36. Rome: Food and Agriculture Organization of the United Nations.

FAO. (in preparation). *Spatial targeting for trypanosomia-sis control in Uganda*, by T.P Robinson, S. Massawe & A. Mugenyi. Pro-Poor Livestock Policy Initiative Working Paper (draft). Rome: Food and Agriculture Organization of the United Nations.

Fischer, G., van Velthuizen, H., Shah, M. & Nachtergaele, F. (2002). *Global Agro-Ecological Assessment for Agriculture in the 21st Century: Methodology and Results*. Research Report RR-02-02. Laxenburg: International Institute for Applied Systems Analysis and the Food and Agriculture Organization of the United Nations.

Freier, J.E., Miller, R.S. & Geter, K.D. (2007). Geospatial analysis and modelling in the prevention and control of animal diseases in the United States. *Veterinaria Italiana*, 43, 549–557.

Gerber, P., Chilonda, P., Franceschini, G. & Menzi, H. (2005). Geographical determinants and environmental implications of livestock production intensification in Asia. *Bioresource Technology* 96, 263–276.

Gilbert, M., Grégoire, J.-C., Freise, J. & Heitland, W. (2004). Long-distance dispersal and human popula-tion density allow the prediction of invasive patterns in the horse-chestnut leafminer *Cameraria ohridella*. *Journal of Animal Ecology* 73, 459-468.

Gilbert, M., Mitchell, A., Bourn, D., Mawdsley, J., Clifton-Hadley, R. & Wint, G.R.W. (2005). The role of cattle movements in determining the distribution and spread of Bovine Tuberculosis in Britain. *Nature* 435, 491-496.

Goetz, S.J., Prince, S.D. & Small, J. (2000). Advances in satellite remote sensing of environmental variables for epidemiological applications. *Advances in Parasitology* 47, 289-307.

Government of Kenya (1996). *Kenyan rangelands 1977-1994: Summary of population estimates for wildlife and livestock*. Nairobi: Government of Kenya, Ministry of Planning and National Development, Department of Resource Surveys and Remote Sensing.

Green, R.M. & Hay, S.I. (2002). The potential of Pathfinder AVHRR data for providing surrogates of climatic vari-ables across Africa and Europe for epidemiological applications. *Remote Sensing of Environment* 79, 165-175.

Grigg, D. (1972). *The Agricultural Systems of the World*. Cambridge: Cambridge University Press.

Griguolo, S. & Mazzanti, M. (1996). ADDAPIX: pixel-by-pixel classification for zoning and monitoring. Rome: FAO Technical Report SD:GCP/INT/578/NET.

Hay, S.I. (2000). An overview of remote sensing and geod-esy for epidemiology and public health application. *Advances in Parasitology* 47, 1-35.

Hay, S.I. & Lennon, J.J. (1999). Deriving meteorological variables across Africa for the study and control of vector-borne disease: a comparison of remote sensing and spatial interpolation of climate. *Tropical Medicine and International Health* 4, 58-71.

Hay, S.I., Tucker, C.J., Rogers, D.J.R. & Packer, M.J. (1996). Remotely sensed surrogates of meteorological data for the study of the distribution and abundance of arthropod vectors of disease. *Annals of Tropical Medicine and Parasitology* 90, 1-19.

Hay, S.I., Cox, J., Rogers, D.J., Randolph, S.E., Stern, D.I., Shanks, G.D., Myers, M.F. & Snow, R.W. (2002). Regional warming and malaria resurgence. *Nature* 420, 628.

Hay, S.I., Randolph, S.E. & Rogers, D.J. (eds) (2000). *Remote Sensing and Geographical Information Systems for Epidemiology*. San Diego: Academic Press.

Hay, S.I., Tatem, A.J., Graham, A.J., Goetz, S.J. & Rogers, D.J. (2006). Global environmental data for mapping infectious disease distribution. *Advances in Parasitology* 62, 37-77.

Hyman, G., Nelson, A. & Lema, G. (2000). *Latin America and Caribbean Population Database 2000*. International Center for Tropical Agriculture (CIAT), Cali, Colombia, as a cooperative activity between UNEP/GRID, CIAT and WRI.

Jahnke, H.E. (1982). *Livestock production systems and livestock development in tropical Africa*. Kiel: Kieler Wisenschaftsverlag Vauk. 253 pp.

Jones P.G. & Thornton P.K. (1999). Fitting a third-order Markov rainfall model to interpolated climate surfaces. *Agricultural and Forest Meteorology* 97, 213–231.

Loveland, T.R., Reed, B.C., Brown, J.F., Ohlen, D.O., Zhu, Z., Yang, L. & Merchant, J. (2000). Development of global land cover characteristics database and IGBP DISCover from 1 km AVHRR data. *International Journal of Remote Sensing* 21, 1303–1330.

McPherson, J.A., Jetz, W. & Rogers, D.J. (2006). Using coarse-grained occurrence data to predict species distributions at finer spatial resolutions - possibilities and limitations. *Ecological Modelling* 192, 499-522.

Norton-Griffiths, M. (1978). *Counting animals: a series of handbooks on techniques in African wildlife ecology.* Handbook no. 1. Nairobi: African Wildlife Leadership Foundation.

Otte, J., Chilonda, P., Slingenbergh, J. & Wint, W. (2001). *The use of geographical information system in the quantitative characterization of livestock production in sub-Saharan Africa.* Poster presentation. Society for Veterinary Epidemiology and Preventive Veterinary Medicine Annual Conference, Noordwijkerhout, The Netherlands, 28-30 March 2001.

Price, J.C. (1984). Land surface temperature measurement for the split window channels of the NOAA 7 Advanced Very High Resolution Radiometer. *Journal of Geophysical Research* 8, 7231-7237.

Putt, S.N.H., Shaw, A.P.M., Matthewman, R.W., Bourn D.M., Underwood, M., James, A.D., Hallam, M.J. & Ellis, P.R. (1980). *The social and economic implications of trypanosomosis control - a study of its impact on livestock production and rural development in northern Nigeria.* Study no. 25. pp 549. Reading: Veterinary Epidemiology and Economics Research Unit.

Reid, R.S., Kruska R.L., Deichmann. U., Thornton, P.K. & Leak, S.G.A. (2000). Will human population growth and land-use change control tsetse during our lifetimes? *Agriculture, Ecosystems and Environment* 77, 227-236.

Robinson, T.P., Harris, R.S., Hopkins, J.S. & Williams, B. G. (2002). An example of decision support for trypanosomiasis control using a Geographic Information System in eastern Zambia. *International Journal of Geographical Information Systems* 16, 345-360.

Robinson, T.P., Franceschini, G. & Wint, G.R.W. (2007). FAO's gridded livestock of the world. *Veterinaria Italiana*, 43, 745–751.

Rogers, D.J. & Williams, B.G. (1994). *Tsetse distribution in Africa: seeing the wood and the trees.* In Edwards, P.J., May, R.M. & Webb., N.R. (eds). *Large-scale ecology and conservation biology.* 35th symposium of the British Ecological Society with the Society for Conservation Biology. Chapter 11 pp. 249-273. Oxford: Blackwell Scientific Publications.

Rogers, D.J. (1997). Satellite Imagery and the Prediction of Tsetse Distributions in East Africa. In *Diagnosis and Control of Livestock Diseases using Nuclear and related techniques.* pp. 397-420. Vienna: International Atomic Energy Agency.

Rogers, D.J. (2000). Satellites, space, time and the African trypanosomiases. *Advances in Parasitology* 47, 129-171.

Rogers, D.J. & Robinson, T.P. (2004). Tsetse Distribution. In: Maudlin, I. Holmes, P. & Miles, M (eds). *The Trypanosomiaises.* pp 624. Wallingford: CABI Publishing.

Rogers, D.J., Hay, S.I. & Packer, M.J. (1996). Predicting the distribution of tsetse flies in West Africa using temporal Fourier processed meteorological satellite data. *Annals of Tropical Medicine and Parasitology* 90, 225-241.

Ruthenberg, H. (1980). *Farming Systems in the Tropics,* Third Edition. Oxford: Clarendon Press.

Shaw, A.P.M. (1986). *The control of trypanosomosis in the Sudan and northern Guinea zones of West Africa, with special reference to Nigeria and Mali.* pp. 338. PhD thesis, University of Reading..

Shaw, A.P.M., Hendrickx, G., Gilbert, M., Mattioli, R., Codjia, V., Dao, B., Diall, O., Mahama, C., Sidibé, I. & Wint, W. (2006). *Mapping the benefits: a decision tool for tsetse and trypanosomosis interventions.* Research Report. Department for International Development, Animal Health Programme, Centre for Tropical Veterinary Medicine, University of Edinburgh, and Programme Against African Trypanosomiasis, Food and Agriculture Organization of the United Nations.

Skidmore, A. (ed.) (2002). *Environmental modelling with GIS and remote sensing.* London, New York: Taylor and Francis.

Thornton, P.K., Kruska, R.L., Henninger, N., Kristjanson, P.M., Reid, R.S., Atieno, F., Odero, A. & Ndegwa, T. (2002). *Mapping poverty and livestock in the developing world*. pp. 124. Nairobi: International Livestock Research Institute.

Tucker, C.J. & Sellers, P.J. (1986). Satellite remote sensing of primary production. *International Journal of Remote Sensing* 7, 1395-1416 .

White, D.H. (1998). *A global agro-climatic analysis of the distribution and production of livestock commodities*. Economic Evaluation Unit Working Paper No. 30. Canberra: Australian Centre for Agricultural Research.

Wint, G.R.W., Robinson, T.P., Bourn, D.M., Durr, P.A., Hay, S.I., Randolph, S.E. & Rogers, D.J. (2002). Mapping bovine tuberculosis in Great Britain using environmental data. *Trends in Microbiology* 10, 441-444.

Wint, W. (1996a). *Livestock Geography: A Demonstration of GIS Techniques applied to Global Livestock Systems, and Populations*. Consultancy Report prepared by Environmental Research Group Oxford Ltd. Rome: Animal Production and Health Division, Food and Agriculture Organization of the United Nations.

Wint, W. (1996b). *Livestock Geography II: A further Demonstration of GIS Techniques to Global Livestock Systems, Populations and Productivity*. Consultancy Report prepared by Environmental Research Group Oxford Ltd. Rome: Animal Production and Health Division, Food and Agriculture Organization of the United Nations.

Wint, W. (2003). *Ruminants, seasons and grazing in the Middle East*. Consultancy Report prepared by Environmental Research Group Oxford Ltd. Rome: Animal Production and Health Division, Food and Agriculture Organization of the United Nations.

Wint, W. & Gilbert, M. (2000). Botswana Aerial Survey Information System. Software and Manual produced by Environmental Research Group Oxford Ltd on behalf of the Environment and Development Group, Oxford, for the Department of Wildlife and National Parks, Botswana.

Wint, W., Slingenbergh, J. & Rogers, D. (1999). *Agroecological zones, farming systems and land pressure in Africa and Asia*. Rome. Consultancy Report prepared by Environmental Research Group Oxford Ltd. Rome: Animal Production and Health Division, Food and Agriculture Organization of the United Nations.

Wint, W. & Rogers, D. (1998). *Prediction of cattle density, cultivation levels and farming systems in Kenya*. Consultancy Report prepared by ERGO Ltd and the TALA Research Group, Department of Zoology, University of Oxford. Rome: Animal Production and Health Division, Food and Agriculture Organization of the United Nations., to the Animal Production and Health Division, FAO.

Wint, W. & Slingenbergh, J. (2004). *Global trends in livestock production and epidemiological instability*. Consultancy Report prepared by Environmental Research Group Oxford Ltd. Rome: Animal Production and Health Division, Food and Agriculture Organization of the United Nations.

Wint, W. & Sumption, K. (2005). *Mapping the FMD homelands: An exploratory look at global ruminant production systems associated animal movements, and FMD risk*. Consultancy Report prepared by Environmental Research Group Oxford Ltd. Rome: Animal Production and Health Division, Food and Agriculture Organization of the United Nations.

Wint, W., Rogers, D.J. & Robinson, T.P. (1997). *Ecozones, farming systems and priority areas for tsetse control in East, West and Southern Africa*. Consultancy Report by ERGO Ltd and the TALA Research Group, Department of Zoology, University of Oxford. Rome: Animal Production and Health Division, Food and Agriculture Organization of the United Nations.

Wood, S., Sebastian, K. & Scherr, S.J. (2000). *Pilot analysis of global ecosystems: Agroecosystems*. Joint study by International Food Policy Research Institute (IFPRI) and the World Resources Institute. Washington: IFPRI.

GLOBAL LIVESTOCK NUMBERS, BY COUNTRY AND PRODUCTION SYSTEM.

These appendices provide detailed tables of livestock numbers by country and, where available, by livestock production system. The figures have been estimated by converting the modelled livestock densities into absolute numbers, adjusting them so that they match the national totals given in FAOSTAT for the year 2005 and, where they exist, summarizing them by livestock production system. Livestock data are given for any country and species for which national totals are available in FAOSTAT for 2005.

The livestock production systems used are those of Thornton *et al.*, 2002, as described in Table 7.1 on page 52. At the beginning of Appendices A, B and C a map of the livestock production systems is given for three broad regions. The systems abbreviations provide the column titles in the data tables in each of these appendices.

In these tables, a blank indicates that the system does not occur in a country, whereas a zero indicates that the system exists but does not contain livestock of that species. Where the table entry is "n.a.", either the systems definitions are not available or the livestock data have not been modelled for that country. In either case it is not possible to disaggregate the national livestock figures by livestock production system (so only a national total is given).

Since the Thornton *et al.*, 2002, livestock production system maps were only produced for the developing world, there are many countries for which it is not possible to disaggregate the national livestock figures by livestock production system. For these countries the national totals (from FAOSTAT 2005) of the main livestock species are given in Appendix D.

APPENDIX A

LIVESTOCK NUMBERS FOR CENTRAL AMERICA,
THE CARIBBEAN AND SOUTH AMERICA

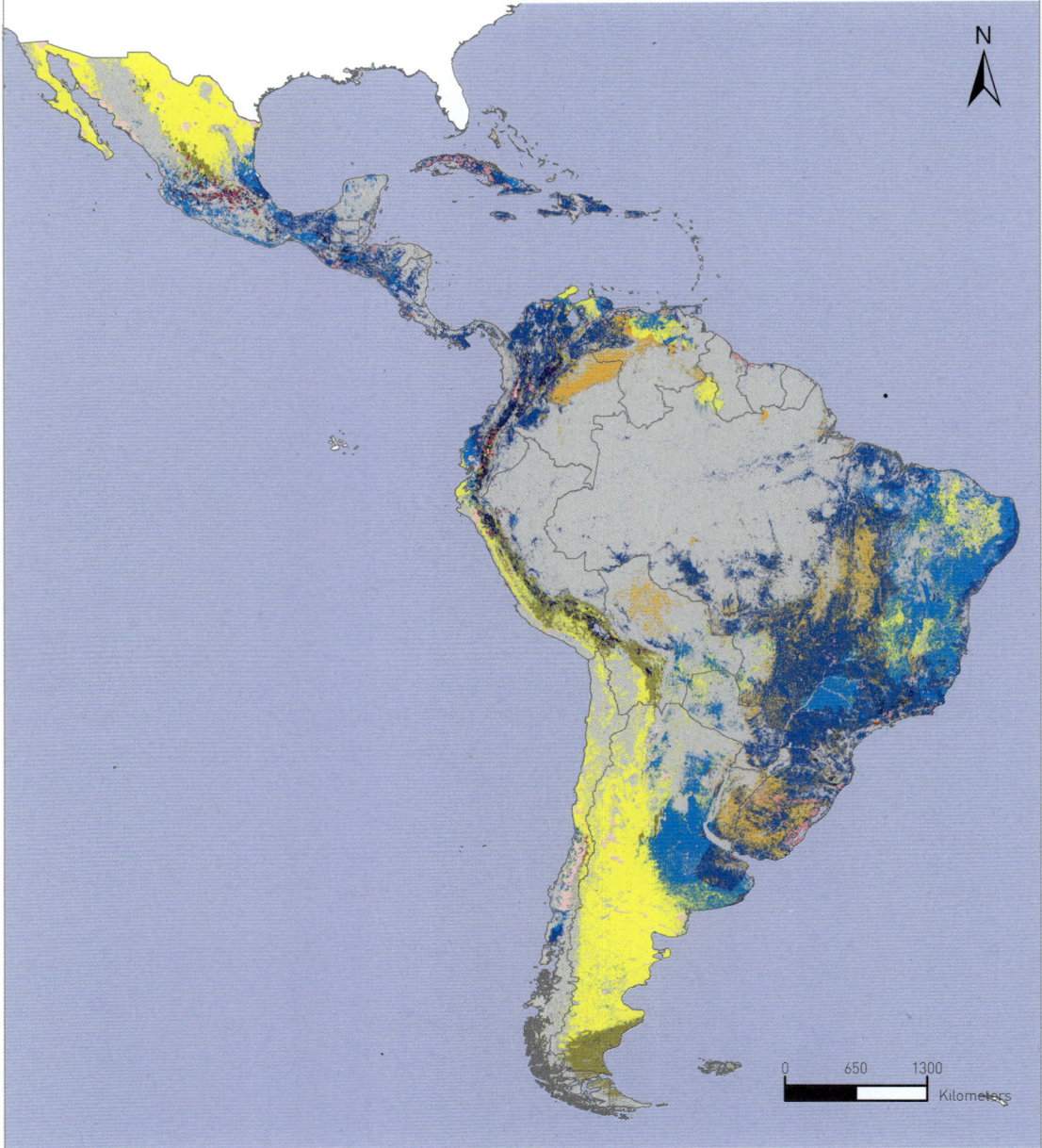

A1 LIVESTOCK PRODUCTION SYSTEMS IN CENTRAL AMERICA, THE CARIBBEAN AND SOUTH AMERICA

Production systems

LGA	MIA	MRA	Other	International boundary
LGH	MIH	MRH	Urban	Water
LGT	MIT	MRT		

Source: Reproduced from Thornton *et al.*, 2002.

TABLE A.1 CATTLE NUMBERS BY COUNTRY AND LIVESTOCK PRODUCTION SYSTEM

Region/Country	LGA	LGH	LGT	MIA	MIH	MIT	MRA	MRH	MRT	OTHER	URBAN	FAOSTAT 2005
CENTRAL AMERICA												
Belize		1 850						14 480		41 470		57 800
Costa Rica		2 300			38 150	8 100	0	207 800	8 450	735 200	0	1 000 000
El Salvador					70 500		13 100	732 900	20 160	411 000	11 550	1 259 210
Guatemala	250	100	400	10 300	247 800	450	160 000	1 060 000	215 700	844 500	500	2 540 000
Honduras	14 500	31 000	4 700	0	67 000		193 320	923 000	107 500	1 159 000	0	2 500 020
Mexico	6 690 000	11 600	842 000	1 697 000	4 900	520 000	5 131 000	3 428 000	1 181 000	12 260 000	34 500	31 800 000
Nicaragua	1 240	7 940			41 550		379 000	1 741 000	32 770	1 285 500	0	3 500 000
Panama		11 400			1 050		0	1 043 750	7 520	536 000	280	1 600 000
CARIBBEAN												
Antigua and Barbuda	n.a.	n.a.	n.a.	n.a.	n.a.	n.a.	n.a.	n.a.	n.a.	n.a.	n.a.	14.300
Bahamas	90	100					0	0		560	0	750
Barbados	n.a.	n.a.	n.a.	n.a.	n.a.	n.a.	n.a.	n.a.	n.a.	n.a.	n.a.	10.300
British Virgin Islands	n.a.	n.a.	n.a.	n.a.	n.a.	n.a.	n.a.	n.a.	n.a.	n.a.	n.a.	2.400
Cayman Islands	n.a.	n.a.	n.a.	n.a.	n.a.	n.a.	n.a.	n.a.	n.a.	n.a.	n.a.	1.300
Cuba	26 740	13 250		176 845	1 218 245		565 040	1 549 050	2 510	392 860	5 460	3 950 000
Dominica	0	13 400						0		0		13 400
Dominican Republic	18 010			150 500	127 550	13 350	122 340	1 317 320	32 770	416 900	1 260	2 200 000
Grenada	0						0	0		4 450		4 450
Guadeloupe		15 250					0	58 250	0	0	0	73 500
Haiti	2 060	2 450	2 400	61 590	51 690	0	95 680	1 046 000	35 430	158 700	0	1 456 000
Jamaica	0			3 800	9 500		9 200	261 800	0	145 700	0	430 000
Martinique	n.a.	n.a.	n.a.	n.a.	n.a.	n.a.	n.a.	n.a.	n.a.	n.a.	n.a.	25 000

(Continued)

TABLE A.1 CATTLE NUMBERS BY COUNTRY AND LIVESTOCK PRODUCTION SYSTEM *(Continued)*

Region/Country	LGA	LGH	LGT	MIA	MIH	MIT	MRA	MRH	MRT	OTHER	URBAN	FAOSTAT 2005
CARIBBEAN *(Continued)*												
Montserrat	n.a.	n.a.	n.a.	n.a.	n.a.	n.a.	n.a.	n.a.	n.a.	n.a.	n.a.	9.700
Netherlands Antilles	0						300			300		600
Puerto Rico	0			0	10 600		14 700	261 600	2 600	130 500	0	420 000
Saint Kitts and Nevis	n.a.	n.a.	n.a.	n.a.	n.a.	n.a.	n.a.	n.a.	n.a.	n.a.	n.a.	4 800
Saint Vincent and the Grenadines	n.a.	n.a.	n.a.	n.a.	n.a.	n.a.	n.a.	n.a.	n.a.	n.a.	n.a.	5 000
Trinidad and Tobago	80							11 170		17 750		29 000
United States Virgin Islands	n.a.	n.a.	n.a.	n.a.	n.a.	n.a.	n.a.	n.a.	n.a.	n.a.	n.a.	8 000
SOUTH AMERICA												
Argentina	13 515 230	3 095 820	120 880	303 480	640	10	15 016 220	7 133 620	10 600	11 568 440	3 060	50 768 000
Bolivia	592 250	1 621 040	612 570	4 900		11 700	370 750	73 270	73 450	3 461 470	800	6 822 200
Brazil	7 208 080	30 340 800	402 320	396 320	3 462 460	16 100	35 695 030	99 955 910	1 108 670	28 412 380	1 930	207 000 000
Chile	447 850	12 420	94 490	1 422 020	8 900	26 180	684 040	832 630	11 460	660 010	0	4 200 000
Colombia	198 500	3 375 850	476 770	173 020	768 240	106 610	474 990	11 850 860	2 708 100	4 867 060	0	25 000 000
Ecuador	246 720	27 220	349 230	214 660	157 670	363 170	1 226 710	915 070	891 720	559 220	0	4 951 390
Falkland Islands (Malvinas)	n.a.	n.a.	n.a.	n.a.	n.a.	n.a.	n.a.	n.a.	n.a.	n.a.	n.a.	4.200
French Guiana		510					0	1 030		7 660		9 200
Guyana	12 660	1 040	1 030	0	27 450		580	1 050		66 190		110 000
Paraguay	179 920	1 108 760	0		31 540		435 770	2 993 530		4 872 820		9 622 340
Peru	632 190	14 980	1 642 880	168 570	1 990	100 240	157 120	140 390	953 340	1 288 150	150	5 100 000
Suriname		1 420			1 420		0	6 180		127 980	0	137 000
Uruguay	5 697 420				433 250			4 641 760		1 227 570	0	12 000 000
Venezuela	2 220 420	3 249 500	139 250	40 090	137 720	49 180	1 534 290	5 566 630	136 300	3 226 370	250	16 300 000

TABLE A.2 NUMBERS OF BUFFALOES BY COUNTRY AND LIVESTOCK PRODUCTION SYSTEM

Region/Country	LGA	LGH	LGT	MIA	MIH	MIT	MRA	MRH	MRT	OTHER	URBAN	FAOSTAT 2005
SOUTH AMERICA												
Brazil	n.a.	n.a.	n.a.	n.a.	n.a.	n.a.	n.a.	n.a.	n.a.	n.a.	n.a.	1 095 000
Suriname	n.a.	n.a.	n.a.	n.a.	n.a.	n.a.	n.a.	n.a.	n.a.	n.a.	n.a.	395
CARIBBEAN												
Trinidad and Tobago	n.a.	n.a.	n.a.	n.a.	n.a.	n.a.	n.a.	n.a.	n.a.	n.a.	n.a.	5 700

TABLE A.3 NUMBERS OF GOATS BY COUNTRY AND LIVESTOCK PRODUCTION SYSTEM

Region/Country	LGA	LGH	LGT	MIA	MIH	MIT	MRA	MRH	MRT	OTHER	URBAN	FAOSTAT 2005
CENTRAL AMERICA												
Belize	0							20	0	145		165
Costa Rica		350			10	0	0	1 300	0	3 040		4 700
El Salvador					270		110	1 910	50	8 380	30	10 750
Guatemala	50	60	330	320	6 790	20	8 920	26 010	20 710	48 710	80	112 000
Honduras	90	300	50	0	230		1 310	8 060	1 220	12 950	0	24 210
Mexico	1 902 740	1 120	581 190	311 070	5 830	286 390	1 173 460	109 880	486 480	4 129 640	3 950	8 991 750
Nicaragua	10	60		20	20		1 090	2 960	80	2 870	0	7 100
Panama	20	20			0		0	1 110	20	5 140	10	6 300
CARIBBEAN												
Antigua and Barbuda	n.a.	n.a.	n.a.	n.a.	n.a.	n.a.	n.a.	n.a.	n.a.	n.a.	n.a.	36 000
Bahamas	1 660	190					0	0		12 650	0	14 500
Barbados	n.a.	n.a.	n.a.	n.a.	n.a.	n.a.	n.a.	n.a.	n.a.	n.a.	n.a.	5 100
British Virgin Islands	n.a.	n.a.	n.a.	n.a.	n.a.	n.a.	n.a.	n.a.	n.a.	n.a.	n.a.	10 000
Cayman Islands	n.a.	n.a.	n.a.	n.a.	n.a.	n.a.	n.a.	n.a.	n.a.	n.a.	n.a.	269
Cuba	7 760	2 740		62 130	191 780		204 780	254 090	133 170	187 390	1 160	1 045 000
Dominica	0	9 700						0		0		9 700
Dominican Republic	1 260	0		20 920	4 360	1 860	18 690	91 650	7 980	43 210	70	190 000
Grenada	0							0		7 200		7 200
Guadeloupe		32 020					0	12 980	0	0		45 000
Haiti	6 050	0	1 350	106 170	47 800	0	198 420	1 251 050	48 040	241 120	n.a.	1 900 000
Jamaica	0			4 560	27 550		1 500	269 970	0	136 420	0	440 000
Martinique	n.a.	n.a.	n.a.	n.a.	n.a.	n.a.	n.a.	n.a.	n.a.	n.a.	n.a.	13 500
Montserrat	n.a.	n.a.	n.a.	n.a.	n.a.	n.a.	n.a.	n.a.	n.a.	n.a.	n.a.	7 000
Netherlands Antilles	n.a.	n.a.	n.a.	n.a.	n.a.	n.a.	n.a.	n.a.	n.a.	n.a.	n.a.	13 500

(Continued)

TABLE A.3 NUMBERS OF GOATS BY COUNTRY AND LIVESTOCK PRODUCTION SYSTEM *(Continued)*

Region/Country	LGA	LGH	LGT	MIA	MIH	MIT	MRA	MRH	MRT	OTHER	URBAN	FAOSTAT 2005
CARIBBEAN *(Continued)*												
Puerto Rico	0			0	470	0	10	7 560	0	960	0	9 000
Saint Kitts and Nevis	n.a.	n.a.	n.a.	n.a.	n.a.	n.a.	n.a.	n.a.	n.a.	n.a.	n.a.	16 000
Saint Vincent and the Grenadines	n.a.	n.a.	n.a.	n.a.	n.a.	n.a.	n.a.	n.a.	n.a.	n.a.	n.a.	7 200
Trinidad and Tobago	1 730							23 750		33 820		59 300
United States Virgin Islands	n.a.	n.a.	n.a.	n.a.	n.a.	n.a.	n.a.	n.a.	n.a.	n.a.	n.a.	4 000
SOUTH AMERICA												
Argentina	2 601 990	13 460	19 500	65 980	0	0	283 210	12 410	960	1 202 110	380	4 200 000
Bolivia	181 640	111 190	238 370	4 370		5 940	93 320	24 760	32 500	808 410	500	1 501 000
Brazil	1 955 850	102 070	3 120	141 170	13 520	120	4 581 340	643 000	14 230	3 245 410	170	10 700 000
Chile	415 940	130	25 490	155 560	330	2 670	35 440	5 030	1 220	93 190	0	735 000
Colombia	10 970	249 400	16 530	4 860	35 680	2 840	28 420	407 880	119 350	324 070	0	1 200 000
Ecuador	11 600	940	12 220	33 860	5 650	13 760	63 270	34 540	43 880	30 280	0	250 000
French Guiana		10					0	20		770		800
Guyana	12 710	2 390	160	0	9 990		620	1 680		51 450		79 000
Paraguay	6 480	10 570			410		13 800	28 870		94 870		155 000
Peru	424 500	570	525 670	107 920	50	22 100	145 910	1 560	379 690	392 030	0	2 000 000
Suriname		90			330		0	480		6 200	0	7 100
Uruguay		7 970			590	1 360		6 080		1 360	0	16 000
Venezuela	218 130	252 730	5 000	3 670	9 580	1 360	141 980	310 720	3 130	373 580	120	1 320 000

TABLE A.4 NUMBERS OF SHEEP BY COUNTRY AND LIVESTOCK PRODUCTION SYSTEM

Region/Country	LGA	LGH	LGT	MIA	MIH	MIT	MRA	MRH	MRT	OTHER	URBAN	FAOSTAT 2005
CENTRAL AMERICA												
Belize	40							1 265		4 960		6 265
Costa Rica		0			120	0	0	590	20	1 970	0	2 700
El Salvador					100		170	570	30	4 220	10	5 100
Guatemala	160	360	1 160	300	7 110	10	17 350	59 890	50 520	122 980	160	260 000
Honduras	40	520	40	0	340		400	4 130	430	8 970	0	14 870
Mexico	1 293 260	700	425 380	252 070	430	541 430	658 880	240 750	560 800	2 838 940	7 130	6 819 770
Nicaragua	0	80		0	0		770	1 500	80	2 070	0	4 500
CARIBBEAN												
Antigua and Barbuda	n.a.	n.a.	n.a.	n.a.	n.a.	n.a.	n.a.	n.a.	n.a.	n.a.	n.a.	19 000
Bahamas	420	140					0	0		5 940	0	6 500
Barbados	n.a.	n.a.	n.a.	n.a.	n.a.	n.a.	n.a.	n.a.	n.a.	n.a.	n.a.	10 800
British Virgin Islands	n.a.	n.a.	n.a.	n.a.	n.a.	n.a.	n.a.	n.a.	n.a.	n.a.	n.a.	6 100
Cuba	17 590	11 160		80 170	789 770		336 980	847 200	21 660	289 350	6 120	2 400 000
Dominica	0	7 600						0		0		7 600
Dominican Republic	830	0		16 530	5 140	7 120	7 460	43 210	5 670	37 010	30	123 000
Grenada	0						0	0	0	13 200		13 200
Guadeloupe		2 910					0	240	0	0		3 150
Haiti	750	0	50	4 230	2 690	0	37 000	94 510	1 880	12 390	0	153 500
Jamaica	0			10	20		0	180	0	1 070	0	1 280
Martinique	n.a.	n.a.	n.a.	n.a.	n.a.	n.a.	n.a.	n.a.	n.a.	n.a.	n.a.	18 000
Montserrat	n.a.	n.a.	n.a.	n.a.	n.a.	n.a.	n.a.	n.a.	n.a.	n.a.	n.a.	4 700
Netherlands Antilles	0						4 340			4 660		9 000
Puerto Rico	0			0	190		230	12 030	0	3 550	0	16 000
Saint Kitts and Nevis	n.a.	n.a.	n.a.	n.a.	n.a.	n.a.	n.a.	n.a.	n.a.	n.a.	n.a.	12 500

(Continued)

TABLE A.4 NUMBERS OF SHEEP BY COUNTRY AND LIVESTOCK PRODUCTION SYSTEM *(Continued)*

Region/Country	LGA	LGH	LGT	MIA	MIH	MIT	MRA	MRH	MRT	OTHER	URBAN	FAOSTAT 2005
CARIBBEAN *(Continued)*												
Saint Vincent and the Grenadines	n.a.	n.a.	n.a.	n.a.	n.a.	n.a.	n.a.	n.a.	n.a.	n.a.	n.a.	12 000
Trinidad and Tobago	0							1 250		2 150		3 400
United States Virgin Islands	n.a.	n.a.	n.a.	n.a.	n.a.	n.a.	n.a.	n.a.	n.a.	n.a.	n.a.	3 200
SOUTH AMERICA												
Argentina	4 420 330	1 551 180	1 057 200	117 730	210	130	862 330	1 161 960	36 370	3 242 560	0	12 450 000
Bolivia	1 017 590	646 730	2 121 340	20 670		36 500	504 500	138 700	211 140	3 851 810	1 020	8 550 000
Brazil	1 698 350	2 634 100	17 850	97 770	755 150	460	4 070 130	2 927 380	39 350	2 959 290	170	15 200 000
Chile	335 690	870	1 152 480	409 130	1 240	3 110	196 110	66 510	55 640	1 179 220	0	3 400 000
Colombia	13 250	449 030	51 990	12 580	51 230	3 690	50 870	756 320	160 390	630 650	0	2 180 000
Ecuador	168 550	8 440	181 960	176 720	32 570	199 070	639 090	307 550	536 930	299 120	0	2 550 000
Falkland Islands (Malvinas)	0		109 600						0	580 400		690 000
French Guiana		20						50		2 530		2 600
Guyana	17 570	3 480	120	0	16 720		1 060	2 330		88 720		130 000
Paraguay	7 240	62 110			2 350		12 390	137 310		278 600		500 000
Peru	780 070	46 440	5 224 000	123 200	310	87 140	124 850	27 500	3 200 570	4 385 640	280	14 000 000
Suriname		50			330		0	330		6 990	0	7 700
Uruguay		5 174 530			288 730			3 479 510		769 230	0	9 712 000
Venezuela	83 750	97 180	9 560	1 950	2 670	340	57 250	112 040	1 130	164 080	50	530 000

TABLE A.5 NUMBERS OF PIGS BY COUNTRY AND LIVESTOCK PRODUCTION SYSTEM

Region/Country	LGA	LGH	LGT	MIA	MIH	MIT	MRA	MRH	MRT	OTHER	URBAN	FAOSTAT 2005
CENTRAL AMERICA												
Belize		300						4 050		16 880		21 230
Costa Rica		970			17 890	210	0	97 400	4 200	429 330	0	550 000
El Salvador					14 120		4 270	99 380	2 780	65 950	1 520	188 020
Guatemala	90	260	220	910	6 560	120	15 910	48 020	33 290	106 520	100	212 000
Honduras	2 600	7 510	1 430	0	9 340		38 510	168 300	21 390	240 920	0	490 000
Mexico	1 816 920	4 540	543 600	781 880	1 410	659 190	2 107 580	680 860	956 240	7 067 120	5 850	14 625 190
Nicaragua	120	1 210		410	330		9 770	39 990	670	70 500	0	123 000
Panama		5 580			150		50	113 850	1 150	151 220	0	272 000
CARIBBEAN												
Antigua and Barbuda	n.a.	n.a.	n.a.	n.a.	n.a.	n.a.	n.a.	n.a.	n.a.	n.a.	n.a.	2 800
Bahamas	840	110					0	0		4 050	0	5 000
Barbados	n.a.	n.a.	n.a.	n.a.	n.a.	n.a.	n.a.	n.a.	n.a.	n.a.	n.a.	19 000
British Virgin Islands	n.a.	n.a.	n.a.	n.a.	n.a.	n.a.	n.a.	n.a.	n.a.	n.a.	n.a.	1 500
Cayman Islands								399		0		399
Cuba	16 680	9 340		50 770	556 500		208 280	846 890	280	308 010	3 250	2 000 000
Dominica	0	1 730						2 450		820		5 000
Dominican Republic	7 840	0		52 770	31 050	8 920	42 870	256 940	16 200	162 540	870	580 000
Grenada	0							0		2 650		2 650
Guadeloupe		1 370					0	25 890	0	2 740		30 000
Haiti	2 640	17 400	2 860	40 470	32 270	0	80 610	675 210	33 310	115 230		1 000 000
Jamaica	420			2 850	6 130		280	48 670	0	26 650	0	85 000
Martinique	n.a.	n.a.	n.a.	n.a.	n.a.	n.a.	n.a.	n.a.	n.a.	n.a.	n.a.	20 000
Montserrat	n.a.	n.a.	n.a.	n.a.	n.a.	n.a.	n.a.	n.a.	n.a.	n.a.	n.a.	1 100
Netherlands Antilles	0						1 800			700		2 500

(Continued)

TABLE A.5 NUMBERS OF PIGS BY COUNTRY AND LIVESTOCK PRODUCTION SYSTEM *(Continued)*

Region/Country	LGA	LGH	LGT	MIA	MIH	MIT	MRA	MRH	MRT	OTHER	URBAN	FAOSTAT 2005
CARIBBEAN *(Continued)*												
Puerto Rico	1 190			1 080	3 300	10	4 320	54 720	1 140	34 250	0	100 000
Saint Kitts and Nevis	n.a.	n.a.	n.a.	n.a.	n.a.	n.a.	n.a.	n.a.	n.a.	n.a.	n.a.	2 000
Saint Vincent and the Grenadines	0									9 150		9 150
Trinidad and Tobago	0							13 110		29 890		43 000
United States Virgin Islands	n.a.	n.a.	n.a.	n.a.	n.a.	n.a.	n.a.	n.a.	n.a.	n.a.	n.a.	2.600
SOUTH AMERICA												
Argentina	210 090	49 790	7 450	8 450	10	10	720 400	140 250	1 080	352 470	0	1 490 000
Bolivia	328 320	159 530	381 270	4 700		13 600	124 940	54 250	50 690	1 865 780	920	2 984 000
Brazil	980 930	2 610 350	149 300	56 280	417 740	17 940	4 466 710	15 742 160	370 520	8 387 560	510	33 200 000
Chile	357 000	7 690	68 070	1 845 470	6 270	159 110	456 680	107 240	10 050	432 420	0	3 450 000
Colombia	16 470	168 450	9 520	4 030	12 750	1 480	24 330	284 790	49 680	578 500	0	1 150 000
Ecuador	54 700	10 740	92 480	183 080	26 370	88 350	454 510	300 310	223 480	514 570	0	1 948 590
Falkland Islands (Malvinas)	0		0						0	30		30
French Guiana		20		0			0	80		10 400		10 500
Guyana	1 020	60	20	0	1 140		30	70		10 660		13 000
Paraguay	2 450	155 240			6 070		3 730	821 660		610 850		1 600 000
Peru	533 680	5 700	707 540	90 220	1 280	52 280	70 480	30 850	378 730	1 029 230	0	2 900 000
Suriname		130			280		0	380		23 710	0	24 500
Uruguay		124 540			9 250			98 630		24 580	0	257 000
Venezuela	190 080	408 250	17 660	20 810	121 490	13 260	150 910	624 380	22 940	1 525 570	4 650	3 100 000

TABLE A.6 NUMBERS OF POULTRY BY COUNTRY AND LIVESTOCK PRODUCTION SYSTEM

Region/Country	LGA	LGH	LGT	MIA	MIH	MIT	MRA	MRH	MRT	OTHER	URBAN	FAOSTAT 2005
CENTRAL AMERICA												
Belize		10						96 290		1 556 700	n.a.	1 653 000
Costa Rica		31 480			732 480	6 570	0	3 850 040	174 000	14 705 430	0	19 500 000
El Salvador					727 020		264 700	7 830 260	214 240	4 079 110	93 670	13 209 000
Guatemala	21 170	18 480	18 150	113 360	1 244 730	10 040	2 134 020	6 408 330	3 793 100	13 212 750	25 870	27 000 000
Honduras	20	2 250	10	0	30		326 770	11 386 050	1 014 800	5 970 070	0	18 700 000
Mexico	49 305 570	50 170	19 236 480	25 415 690	18 460	16 172 510	85 264 950	9 246 390	31 394 750	201 823 910	171 120	438 100 000
Nicaragua	26 660	125 710		64 280	55 730		1 365 020	6 109 280	148 760	10 104 560	0	18 000 000
Panama		402 150			16 640		0	6 114 430	16 410	7 710 370	0	14 260 000
CARIBBEAN												
Antigua and Barbuda	n.a.	n.a.	n.a.	n.a.	n.a.	n.a.	n.a.	n.a.	n.a.	n.a.	n.a.	105 000
Bahamas	349 500	300 160					0	28 210		2 322 130	0	3 000 000
Barbados	n.a.	n.a.	n.a.	n.a.	n.a.	n.a.	n.a.	n.a.	n.a.	n.a.	n.a.	3 450 000
Cayman Islands								6 000		0		6 000
Cuba	135 150	63 980		516 630	7 589 760		1 619 200	11 209 180	48 480	4 723 340	94 280	26 000 000
Dominica	0	63 590						94 810		31 600		190 000
Dominican Republic	398 640	0		2 864 430	3 142 090	761 410	3 404 890	20 840 320	2 198 500	13 752 020	137 700	47 500 000
Grenada	0							610		267 390		268 000
Guadeloupe		17 660					0	430 640	0	54 700		503 000
Haiti	14 410	46 720	20 580	194 820	94 140		488 680	3 874 000	307 300	884 350		5 925 000
Jamaica	80 300			0	118 130		45 710	7 501 720	0	4 754 140	0	12 500 000
Martinique	n.a.	n.a.	n.a.	n.a.	n.a.	n.a.	n.a.	n.a.	n.a.	n.a.	n.a.	590 000
Montserrat	n.a.	n.a.	n.a.	n.a.	n.a.	n.a.	n.a.	n.a.	n.a.	n.a.	n.a.	36 000
Netherlands Antilles	0						10 560			124 440		135 000

(Continued)

TABLE A.6 NUMBERS OF POULTRY BY COUNTRY AND LIVESTOCK PRODUCTION SYSTEM *(Continued)*

Region/Country	LGA	LGH	LGT	MIA	MIH	MIT	MRA	MRH	MRT	OTHER	URBAN	FAOSTAT 2005
CARIBBEAN *(Continued)*												
Puerto Rico	0			0	47 240		339 980	6 302 140	0	4 510 640	0	11 200 000
Saint Kitts and Nevis	n.a.	n.a.	n.a.	n.a.	n.a.	n.a.	n.a.	n.a.	n.a.	n.a.	n.a.	70 000
Saint Vincent and the Grenadines	0									125 000		125 000
Trinidad and Tobago	47 350							8 354 190		19 798 460		28 200 000
United States Virgin Islands	n.a.	n.a.	n.a.	n.a.	n.a.	n.a.	n.a.	n.a.	n.a.	n.a.	n.a.	35 000
SOUTH AMERICA												
Argentina	21 695 190	5 840 200	2 070 940	1 432 230	340	4 150	27 328 000	9 230 730	79 840	32 713 250	130	100 395 000
Bolivia	7 302 250	4 266 650	8 018 100	187 980		207 860	3 494 990	1 714 240	968 810	49 287 870	1 250	75 450 000
Brazil	19 896 460	79 908 390	5 466 970	1 818 410	12 976 590	404 340	144 295 900	616 350 810	14 884 320	223 741 420	6 390	1 119 750 000
Chile	31 750 860	179 010	9 831 590	17 883 560	62 360	1 495 910	5 137 280	2 172 640	1 796 420	51 188 680	1 690	121 500 000
Colombia	899 360	19 722 560	1 886 450	406 990	2 030 160	432 060	1 401 440	31 547 830	11 152 630	55 520 520	0	125 000 000
Ecuador	5 931 230	494 370	3 866 690	10 600 990	2 455 700	5 318 080	23 104 560	13 630 760	12 365 010	26 687 610	0	104 455 000
Falkland Islands (Malvinas)	0		990				0		0	2 010		3 000
French Guiana		230						900		215 870		217 000
Guyana	1 189 810	152 510	9 420	0	2 079 240		115 340	128 160	16 325 520			20 000 000
Paraguay	14 430	1 793 280			69 000		25 650	8 685 220	7 342 420			17 930 000
Peru	15 156 110	259 190	15 999 000	3 447 610	42 410	1 364 150	2 744 300	1 038 070	9 758 580	45 189 320	1 260	95 000 000
Suriname		8 540			73 560		0	45 940		3 737 960	0	3 866 000
Uruguay		6 469 060			558 260			5 450 840		1 886 840	0	14 365 000
Venezuela	12 534 890	15 248 410	784 500	213 050	660 450	91 980	8 283 800	17 572 120	497 460	54 108 040	5 300	110 000 000

APPENDIX B

LIVESTOCK NUMBERS FOR AFRICA

B1 LIVESTOCK PRODUCTION SYSTEMS IN AFRICA

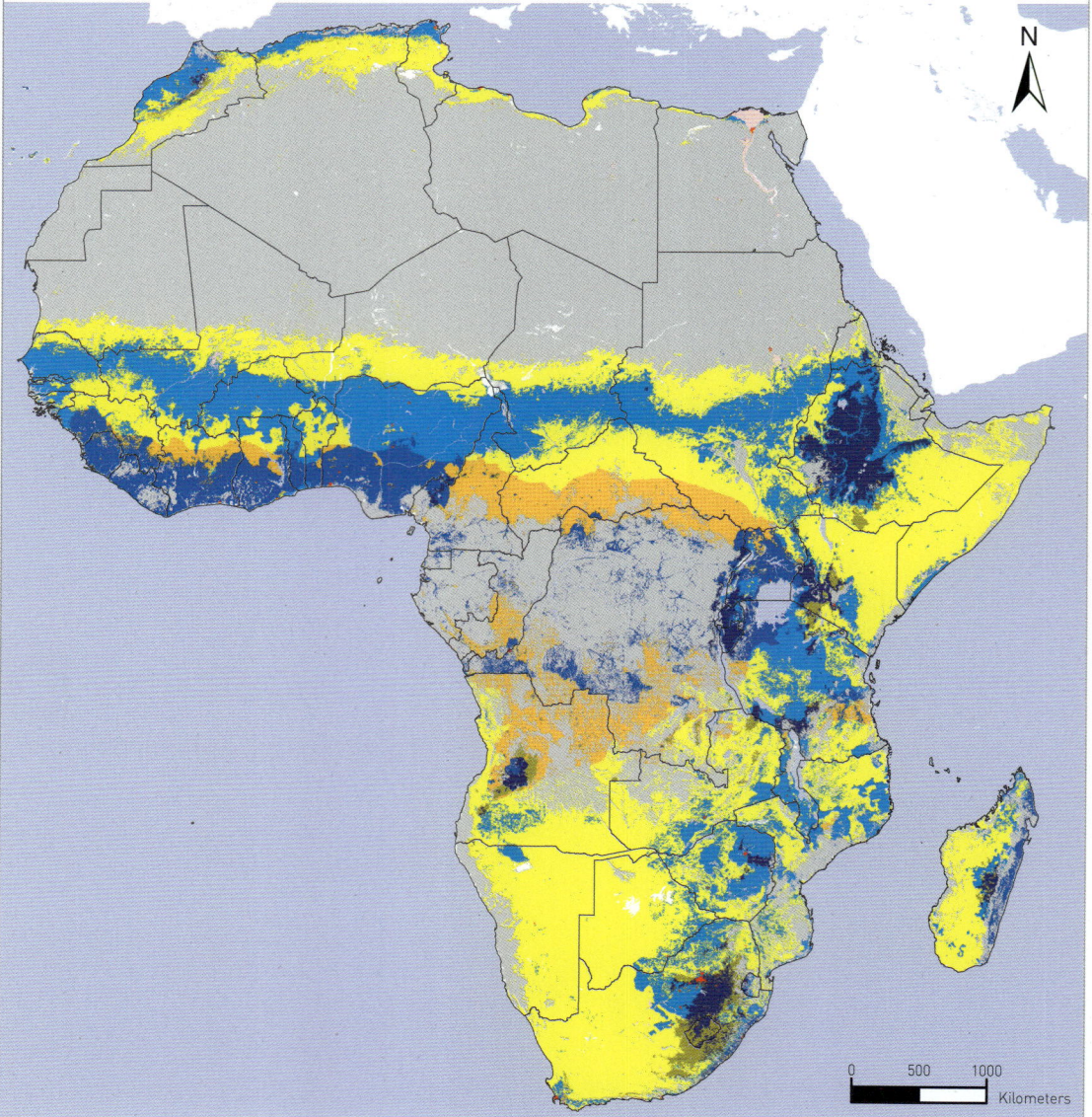

Production systems

LGA	MIA	MRA	Other	International boundary
LGH	MIH	MRH	Urban	Water
LGT	MIT	MRT		

Source: Reproduced from Thornton *et al.*, 2002.

TABLE B.1 CATTLE NUMBERS BY COUNTRY AND LIVESTOCK PRODUCTION SYSTEM

Region/Country	LGA	LGH	LGT	MIA	MIH	MIT	MRA	MRH	MRT	OTHER	URBAN	FAOSTAT 2005
NORTHERN AFRICA												
Algeria	478 170		24 380				442 650	57 650	16 120	536 970	4 060	1 560 000
Egypt	263 800			3 006 180			379 960			837 200	12 860	4 500 000
Libyan Arab Jamahiriya	101 410			0			12 680			15 460	450	130 000
Morocco	701 800	3 110	67 530	5 250		2 740	1 495 270	36 950	37 310	367 570	11 270	2 728 800
Sudan	16 443 100	967 500	6 320	400 150			17 751 500	8 230	20 000	2 719 200	9 000	38 325 000
Tunisia	244 050		320				369 180	12 100		123 620	730	750 000
WESTERN AFRICA												
Benin	960 140	18 000			1 060		654 390	158 270		8 140		1 800 000
Burkina Faso	1 265 550	0		4 710			6 727 240			11 550	1 110	8 010 160
Côte d'Ivoire	665 750	190 200			550		481 980	153 070	0	8 450	0	1 500 000
Gambia	12 870			18 710			284 580			13 840		330 000
Ghana	275 940	115 180					705 960	276 580		11 340	0	1 385 000
Guinea	1 425 270	176 210	40	4 250	1 560		346 320	1 346 700	1 670	95 710	2 270	3 400 000
Guinea-Bissau	198 680			16 400			266 860			48 060		530 000
Liberia								27 800		8 200		36 000
Mali	2 608 250			145 700			4 685 030			261 020	0	7 700 000
Mauritania	891 810			3 230			683 070			21 890	0	1 600 000
Niger	1 622 910			1 650			555 300			79 860	280	2 260 000
Nigeria	1 124 370	146 820	210	40 920	80		11 848 870	1 789 800	161 570	82 550	4 810	15 200 000
Senegal	700 350			22 040			2 317 540			29 950	120	3 070 000
Sierra Leone		12 650		0				358 750	0	28 600		400 000
Togo	8 620						213 330	57 440		610	0	280 000
Cape Verde	n.a.	n.a.	n.a.	n.a.	n.a.	n.a.	n.a.	n.a.	n.a.	n.a.	n.a.	23 000

(Continued)

TABLE B.1 CATTLE NUMBERS BY COUNTRY AND LIVESTOCK PRODUCTION SYSTEM *(Continued)*

Region/Country	LGA	LGH	LGT	MIA	MIH	MIT	MRA	MRH	MRT	OTHER	URBAN	FAOSTAT 2005
MIDDLE AFRICA												
Angola	2 568 180	162 630	218 900				698 490	45 970	184 950	269 830	1 050	4 150 000
Cameroon	351 750	2 265 960	345 150	12 330	100		1 703 360	629 050	521 200	171 000	80	6 000 000
Central African Republic	558 250	2 602 140					3 120	126 240		133 250		3 423 000
Chad	2 987 210			6 830			3 435 200			110 760	0	6 540 000
Congo		53 370						12 690		48 810	130	115 000
Democratic Republic of the Congo	25 640	150 200	2 440				16 470	136 950	119 220	305 880	140	756 940
Equatorial Guinea	10	50	0		0		0	1 110	0	3 880		5 050
Gabon		9 560					0	3 030		22 410	0	35 000
Sao Tome and Principe							1 060	0		3 540		4 600
EASTERN AFRICA												
Burundi							5 210	72 030	234 845	12 915		325 000
Comoros								11 420		33 580		45 000
Djibouti	137 070						8 760			151 170	0	297 000
Eritrea	861 950		21 960	610			693 880		175 580	196 020		1 950 000
Ethiopia	3 743 525	43 625	220 300	7 150			8 735 260	1 280 250	23 198 000	1 271 890	0	38 500 000
Kenya	2 932 925		512 775				2 044 045	1 006 430	4 030 505	1 459 850	13 470	12 000 000
Madagascar	6 272 965	16 465	180 210				1 795 925	463 985	456 865	1 312 720	865	10 500 000
Malawi	33 095	8 900	9 580				584 120	19 800	16 105	77 380	1 020	750 000
Mozambique	502 450	160	3 165				441 650	1 190	2 865	367 800	720	1 320 000
Rwanda			0				275 750	109 370	590 180	28 800		1 004 100
Somalia	4 264 535		1 380	79 315			372 780			631 990	0	5 350 000
Uganda	262 500	208 570	10 560				1 536 415	2 791 160	856 155	434 640	0	6 100 000

(Continued)

TABLE B.1 CATTLE NUMBERS BY COUNTRY AND LIVESTOCK PRODUCTION SYSTEM *(Continued)*

Region/Country	LGA	LGH	LGT	MIA	MIH	MIT	MRA	MRH	MRT	OTHER	URBAN	FAOSTAT 2005
EASTERN AFRICA *(Continued)*												
United Republic of Tanzania	2 379 260	105 400	245 900				9 919 300	3 240 575	778 060	1 129 720	1 785	17 800 000
Zambia	1 602 420	3 420	12 740				728 030	170	670	246 110	6 440	2 600 000
Zimbabwe	1 445 340		25 110				3 320 450		292 180	295 640	21 280	5 400 000
SOUTHERN AFRICA												
Botswana	2 547 380						529 050			22 120	1 450	3 100 000
Lesotho	2 000		270 000						266 830		1 170	540 000
Namibia	1 909 090		160				1 146 320		7	64 330	13 980	3 133 887
South Africa	4 171 000	102 540	2 343 120	14 420			3 305 070	497 460	2 159 220	1 107 500	63 670	13 764 000
Swaziland	76 120						135 600	14 650	99 850	251 600	2 180	580 000

TABLE B.2 NUMBERS OF BUFFALOES BY COUNTRY AND LIVESTOCK PRODUCTION SYSTEM

Region/Country	LGA	LGH	LGT	MIA	MIH	MIT	MRA	MRH	MRT	OTHER	URBAN	FAOSTAT 2005
NORTHERN AFRICA												
Egypt	n.a.	n.a.	n.a.	n.a.	n.a.	n.a.	n.a.	n.a.	n.a.	n.a.	n.a.	3 920 000

TABLE B.3 NUMBERS OF GOATS BY COUNTRY AND LIVESTOCK PRODUCTION SYSTEM

Region/Country	LGA	LGH	LGT	MIA	MIH	MIT	MRA	MRH	MRT	OTHER	URBAN	FAOSTAT 2005
NORTHERN AFRICA												
Algeria	1 653 960		43 690				349 350	19 060	12 200	1 112 170	9 570	3 200 000
Egypt	261 050			1 656 050			309 230			1 727 400	6 270	3 960 000
Libyan Arab Jamahiriya	1 092 670			10		5 790	111 140			52 080	9 100	1 265 000
Morocco	2 778 070	3 080	178 760	35 900			1 114 150	38 260	97 620	1 104 890	2 080	5 358 600
Sudan	17 270 460	2 281 790	6 270	403 840			18 808 980	26 170	28 760	3 062 870	110 860	42 000 000
Tunisia	937 300		260				206 810	6 260		243 450	5 920	1 400 000
Western Sahara	110									172 890	0	173 000
WESTERN AFRICA												
Benin	487 890	25 340			4 700		562 600	283 870		15 600		1 380 000
Burkina Faso	1 885 340	520		10 640			8 792 210			19 360	930	10 709 000
Côte d'Ivoire	222 550	127 390			40		130 180	667 810	0	44 030	0	1 192 000
Gambia	22 500			16 410			218 100			12 990		270 000
Ghana	389 410	197 630					1 273 700	1 652 170		118 690	0	3 631 600
Guinea	572 220	90 660	70	560	90		210 420	460 040	150	22 780	4 010	1 361 000
Guinea-Bissau	71 140			19 030			209 330			35 500		335 000
Liberia								163 520		56 480		220 000
Mali	4 105 590			238 190			6 373 970			1 332 250	0	12 050 000
Mauritania	3 479 000			43 820			1 598 950			478 210	20	5 600 000
Niger	4 920 950			4 820			972 030			996 610	5 590	6 900 000
Nigeria	1 414 260	168 410	50	36 160	6 490		14 018 630	11 498 630	46 480	712 860	98 030	28 000 000
Senegal	569 380			46 640			3 448 270			37 040	3 670	4 105 000
Sierra Leone		2 140		0				203 150	0	14 710		220 000
Togo	11 610						969 040	483 660		15 690	0	1 480 000
Cape Verde	n.a.	n.a.	n.a.	n.a.	n.a.	n.a.	n.a.	n.a.	n.a.	n.a.	n.a.	112 750

(Continued)

TABLE B.3 NUMBERS OF GOATS BY COUNTRY AND LIVESTOCK PRODUCTION SYSTEM *(Continued)*

Region/Country	LGA	LGH	LGT	MIA	MIH	MIT	MRA	MRH	MRT	OTHER	URBAN	FAOSTAT 2005
MIDDLE AFRICA												
Angola	1 196 740	180 530	95 530				324 910	5 870	24 520	221 890	10	2 050 000
Cameroon	166 350	1 744 730	356 700	2 490	1 440		586 820	534 180	138 630	868 500	160	4 400 000
Central African Republic	750 730	2 019 220					9 550	98 740		208 760		3 087 000
Chad	3 021 200			8 720			2 503 200			309 480	0	5 842 600
Congo	.	70 490						20 010		203 720	780	295 000
Democratic Republic of the Congo	41 840	853 550	10 170				16 340	1 001 530	289 230	1 805 980	3 280	4 021 920
Equatorial Guinea	0	60	0		0		0	2 100	0	6 840	0	9 000
Gabon		8 690						4 940		76 210	160	90 000
Sao Tome and Principe							1 680	0		3 320		5 000
EASTERN AFRICA												
Burundi							6 960	146 210	559 500	37 330		750 000
Comoros								55 440		59 560		115 000
Djibouti	252 170						39 510			220 320	0	512 000
Eritrea	613 970		20 650	2 510			544 910		349 440	168 520		1 700 000
Ethiopia	1 381 870	6 400	77 950	3 000			4 158 570	111 940	3 606 270	280 000	0	9 626 000
Kenya	5 293 200		652 200				2 772 050	436 420	1 699 560	1 142 740	3 830	12 000 000
Madagascar	1 017 090	70	430				111 200	1 940	240	69 030	0	1 200 000
Malawi	89 380	22 210	25 850				1 402 820	59 260	30 660	269 200	620	1 900 000
Mozambique	159 830	440	570				136 250	750	300	93 510	350	392 000
Rwanda			0	115 040			251 980	172 550	837 400	77 810		1 339 740
Somalia	8 919 270		29 550				569 890			3 066 250	0	12 700 000
Uganda	91 140	123 570	5 540				1 123 220	3 734 340	1 642 740	979 450	0	7 700 000

(Continued)

TABLE B.3 NUMBERS OF GOATS BY COUNTRY AND LIVESTOCK PRODUCTION SYSTEM *(Continued)*

Region/Country	LGA	LGH	LGT	MIA	MIH	MIT	MRA	MRH	MRT	OTHER	URBAN	FAOSTAT 2005
EASTERN AFRICA *(Continued)*												
United Republic of Tanzania	2 088 190	174 320	213 250				5 929 740	2 116 900	606 010	1 421 240	350	12 550 000
Zambia	603 240	7 760	10 260				474 710	3 060	820	168 800	1 350	1 270 000
Zimbabwe	961 830		2 380				1 745 100		59 080	199 990	1 620	2 970 000
SOUTHERN AFRICA												
Botswana	1 441 400						491 350			12 380	4 870	1 950 000
Lesotho	2 390		337 930						308 880		800	650 000
Namibia	1 454 700		860				482 370		60	101 100	4 390	2 043 480
South Africa	2 935 570	66 700	740 020	24 470			1 166 620	421 210	339 500	691 300	21 610	6 407 000
Swaziland	35 900						64 870	7 310	44 640	120 070	1 210	274 000

TABLE B.4 NUMBERS OF SHEEP BY COUNTRY AND LIVESTOCK PRODUCTION SYSTEM

Region/Country	LGA	LGH	LGT	MIA	MIH	MIT	MRA	MRH	MRT	OTHER	URBAN	FAOSTAT 2005
NORTHERN AFRICA												
Algeria	11 686 490		208 730				2 044 540	78 930	59 010	4 599 350	22 950	18 700 000
Egypt	370 770			1 981 930			364 320			2 424 500	8 480	5 150 000
Libyan Arab Jamahiriya	2 406 610			15 280			216 060			1 838 490	23 560	4 500 000
Morocco	5 431 310	10 080	610 360	24 620		18 550	7 369 880	61 370	515 380	2 944 970	39 780	17 026 300
Sudan	20 297 510	2 137 660	5 840	378 160			21 785 490	14 290	31 440	3 322 030	27 580	48 000 000
Tunisia	4 168 650		2 150				1 670 180	31 840		819 430	7 750	6 700 000
Western Sahara	100									33 900	0	34 000
WESTERN AFRICA												
Benin	253 840	11 050			3 710		327 920	148 220		5 260		750 000
Burkina Faso	946 910	170		6 690			6 041 900			12 940	800	7 009 410
Côte d'Ivoire	195 830	103 680			1 290		134 830	994 520	0	92 850	0	1 523 000
Gambia	9 950			10 570			125 530			1 950		148 000
Ghana	251 100	167 170					939 340	1 703 670		149 820	0	3 211 100
Guinea	499 320	93 690	700	1 070	40		109 350	412 180	390	23 260	0	1 140 000
Guinea-Bissau	113 810			5 760			108 250			72 180		300 000
Liberia								166 660		43 340		210 000
Mali	2 764 210			92 900			4 265 640			1 247 250	0	8 370 000
Mauritania	4 268 730			20 740			2 688 980			1 871 060	490	8 850 000
Niger	3 425 600			1 700			532 440			539 530	730	4 500 000
Nigeria	1 377 250	152 410	60	45 470	2 030		14 552 200	6 128 550	124 040	559 790	58 200	23 000 000
Senegal	964 820			118 350			3 702 480			79 760	6 590	4 872 000
Sierra Leone		7 550		0			0	310 000	0	57 450	0	375 000
Togo	12 090						1 189 390	632 130		16 390	0	1 850 000
Cape Verde	n.a.	n.a.	n.a.	n.a.	n.a.	n.a.	n.a.	n.a.	n.a.	n.a.	n.a.	10 000

(Continued)

TABLE B.4 NUMBERS OF SHEEP BY COUNTRY AND LIVESTOCK PRODUCTION SYSTEM *(Continued)*

Region/Country	LGA	LGH	LGT	MIA	MIH	MIT	MRA	MRH	MRT	OTHER	URBAN	FAOSTAT 2005
MIDDLE AFRICA												
Angola	188 320	24 330	33 790				50 000	660	6 700	36 190	10	340 000
Cameroon	193 530	1 761 980	81 800	2 480	2 660		562 170	468 410	40 980	684 890	1 100	3 800 000
Central African Republic	93 740	135 870					2 550	6 360		20 480		259 000
Chad	1 233 700			3 950			1 144 230			246 120	0	2 628 000
Congo		41 200						3 240		54 560	0	99 000
Democratic Republic of the Congo	20 760	153 880	3 660				2 220	177 420	104 310	437 100	220	899 570
Equatorial Guinea	150	720	0		0		0	8 770	0	27 960		37 600
Gabon		25 890						7 740		161 370	0	195 000
Sao Tome and Principe							1 780	0		1 220		3 000
EASTERN AFRICA												
Burundi							600	128 980	98 500	1 920		230 000
Comoros								8 120		12 880		21 000
Djibouti	219 770						6 190			240 040	0	466 000
Eritrea	841 350		15 420	2 650			916 520		117 060	207 000		2 100 000
Ethiopia	1 464 080	13 570	85 350	1 520			3 424 720	344 310	10 851 520	814 930	0	17 000 000
Kenya	3 876 670		802 920				1 729 590	388 820	2 250 870	946 830	4 300	10 000 000
Madagascar	472 820	0	2 820				53 310	3 170	5 730	112 150	0	650 000
Malawi	6 650	750	1 660				80 780	4 480	2 250	18 410	20	115 000
Mozambique	51 160	10	510				45 880	230	30	27 080	100	125 000
Rwanda							28 320	35 280	377 970	22 760		464 330
Somalia	9 389 480		16 530	58 830			407 990			3 227 170	0	13 100 000
Uganda	57 060	32 580	2 870				244 150	461 700	228 360	123 280	0	1 150 000

(Continued)

TABLE B.4 NUMBERS OF SHEEP BY COUNTRY AND LIVESTOCK PRODUCTION SYSTEM *(Continued)*

Region/Country	LGA	LGH	LGT	MIA	MIH	MIT	MRA	MRH	MRT	OTHER	URBAN	FAOSTAT 2005
EASTERN AFRICA *(Continued)*												
United Republic of Tanzania	604 390	13 520	128 630				1 908 150	382 440	201 420	282 440	10	3 521 000
Zambia	111 070	1 840	820				18 470	110	20	17 670	0	150 000
Zimbabwe	189 790		460				319 760		71 730	27 330	930	610 000
SOUTHERN AFRICA												
Botswana	219 850						74 540			5 120	490	300 000
Lesotho	1 700		422 600						425 580		120	850 000
Namibia	2 494 390		700				70 470		150	97 150	940	2 663 800
South Africa	13 856 580	151 920	4 686 920	41 320			2 269 030	216 370	3 391 470	653 730	49 080	25 316 420
Swaziland	5 470						3 900	490	8 770	8 310	60	27 000

TABLE B.5 NUMBERS OF PIGS BY COUNTRY AND LIVESTOCK PRODUCTION SYSTEM

Region/Country	LGA	LGH	LGT	MIA	MIH	MIT	MRA	MRH	MRT	OTHER	URBAN	FAOSTAT 2005
NORTHERN AFRICA												
Algeria	580		10				130	0	0	4 810	170	5 700
Egypt	170			510			100			29 220	0	30 000
Morocco	3 710	0	50	0		0	1 260	10	90	2 870	10	8 000
Tunisia	2 700		10				610	0		2 680	0	6 000
WESTERN AFRICA												
Benin	150 450	5 180			1 670		99 660	62 200		2 840		322 000
Burkina Faso	171 370	1 200		1 870			2 105 780			2 590	850	2 283 660
Côte d'Ivoire	11 710	14 880			30		13 020	270 230	0	35 130	0	345 000
Gambia	1 490			2 960			13 480			1 070		19 000
Ghana	43 880	13 250					94 620	138 470		14 780	0	305 000
Guinea	30 170	5 230	0	40	0		7 250	23 150	10	1 640	10	67 500
Guinea-Bissau	34 400			5 430			72 600			257 570		370 000
Liberia								94 250		35 750		130 000
Mali	17 300			920			49 520			260	0	68 000
Niger	31 130			0			5 600			2 770	0	39 500
Nigeria	50 950	103 120	120	3 130	2 220		2 397 160	3 807 180	10 290	272 570	3 260	6 650 000
Senegal	41 620			5 320			244 570			14 490	0	306 000
Sierra Leone	0	1 020		0				46 170		4 810		52 000
Togo	13 200						141 400	163 800		1 600	0	320 000
Cape Verde	n.a.	n.a.	n.a.	n.a.	n.a.	n.a.	n.a.	n.a.	n.a.	n.a.	n.a.	205 000
MIDDLE AFRICA												
Angola	461 690	51 090	54 210				132 430	2 150	14 220	64 190	20	780 000
Cameroon	71 980	457 920	16 750	550	90		110 460	126 580	22 550	542 980	140	1 350 000
Central African Republic	112 050	570 750					820	81 180		40 200		805 000

(Continued)

TABLE B.5 NUMBERS OF PIGS BY COUNTRY AND LIVESTOCK PRODUCTION SYSTEM *(Continued)*

Region/Country	LGA	LGH	LGT	MIA	MIH	MIT	MRA	MRH	MRT	OTHER	URBAN	FAOSTAT 2005
MIDDLE AFRICA *(Continued)*												
Chad	8 550			180			14 770			1 500	0	25 000
Congo		21 100						4 250		20 880	270	46 500
Democratic Republic of the Congo	5 040	249 790	570				2 510	259 860	24 950	416 140	220	959 080
Equatorial Guinea	0	0	0		0		0	0	0	6 100		6 100
Gabon		11 670						9 580		190 620	130	212 000
Sao Tome and Principe							780	0		1 720		2 500
EASTERN AFRICA												
Burundi							1 850	18 770	46 200	3 180		70 000
Ethiopia	2 470	0	20				7 810	120	17 950	610	20	29 000
Kenya	69 240		6 990				122 620	44 010	31 490	124 090	16 560	415 000
Madagascar	488 950	550	50 150	0			253 470	245 030	128 190	433 460	200	1 600 000
Malawi	16 270	4 110	3 310				313 220	12 790	8 920	97 250	430	456 300
Mozambique	58 830	730	90				83 670	470	110	35 850	250	180 000
Rwanda			0				33 660	121 420	180 750	11 090		346 920
Somalia	2 630		60	80			360			1 040	30	4 200
Uganda	2 770	8 130	140				155 640	887 680	116 970	127 820	850	1 300 000
United Republic of Tanzania	90 740	4 810	4 620				181 390	24 840	52 440	95 960	200	455 000
Zambia	163 040	1 600	870				140 700	180	20	32 340	1 250	340 000
Zimbabwe	183 730		480	1 730			355 820		26 820	39 310	3 840	610 000
SOUTHERN AFRICA												
Botswana	7 240						740			20	0	8 000
Lesotho	150		31 510						33 200		140	65 000
Namibia	18 550		0				8 970		0	430	50	28 000
South Africa	285 030	5 920	321 270				331 400	72 920	507 520	109 940	12 270	1 648 000
Swaziland	3 190						7 520	1 020	6 530	11 500	240	30 000

TABLE B.6 NUMBERS OF POULTRY BY COUNTRY AND LIVESTOCK PRODUCTION SYSTEM

Region/Country	LGA	LGH	LGT	MIA	MIH	MIT	MRA	MRH	MRT	OTHER	URBAN	FAOSTAT 2005
NORTHERN AFRICA												
Algeria	50 899 430		2 066 040				36 087 940	3 011 580	779 400	31 981 580	313 030	125 139 000
Egypt	2 894 000		78 282 130				6 704 200			27 070 860	198 810	115 150 000
Libyan Arab Jamahiriya	13 015 010			1 900		177 260	931 620			10 876 350	175 120	25 000 000
Morocco	40 093 240	130 380	4 583 180	211 930			64 911 580	1 384 350	2 838 540	22 273 000	396 540	137 000 000
Sudan	9 743 640	430 670	8 740	733 370			8 835 910	9 370	34 950	17 131 800	71 550	37 000 000
Tunisia	36 365 600		32 720				21 065 740	353 820		10 541 600	40 520	68 400 000
WESTERN AFRICA												
Benin	4 766 400	195 320			82 100		5 106 600	2 779 450		70 130		13 000 000
Burkina Faso	2 201 990	250		25 900			23 464 740			41 960	4 160	25 739 000
Côte d'Ivoire	5 381 760	7 974 120			57 260		3 351 320	15 305 980	0	929 560	0	33 000 000
Gambia	31 430			68 240			529 760			20 570		650 000
Ghana	1 601 140	996 050					12 769 550	13 939 930		693 330	0	30 000 000
Guinea	6 443 700	1 230 820	5 280	16 630	2 330		1 713 120	5 123 140	4 340	457 150	3 490	15 000 000
Guinea-Bissau	579 130			81 620			732 870			206 380		1 600 000
Liberia								3 976 040		1 523 960		5 500 000
Mali	6 602 220			421 340			22 840 510			1 135 930	0	31 000 000
Mauritania	1 909 640			6 650			326 900			1 917 100	39 710	4 200 000
Niger	11 979 710			23 520			6 121 300			6 864 480	10 990	25 000 000
Nigeria	3 551 220	1 202 870	120	115 220	91 560		42 262 700	82 697 880	239 050	7 053 380	2 786 000	140 000 000
Senegal	3 098 350			851 880			21 978 110			839 280	191 380	26 959 000
Sierra Leone		108 450		0				7 042 750		418 800		7 570 000
Togo	309 920						4 951 460	3 683 220		55 400	0	9 000 000
Cape Verde	n.a.	n.a.	n.a.	n.a.	n.a.	n.a.	n.a.	n.a.	n.a.	n.a.	n.a.	460 000

(Continued)

TABLE B.6 NUMBERS OF POULTRY BY COUNTRY AND LIVESTOCK PRODUCTION SYSTEM *(Continued)*

Region/Country	LGA	LGH	LGT	MIA	MIH	MIT	MRA	MRH	MRT	OTHER	URBAN	FAOSTAT 2005
MIDDLE AFRICA												
Angola	2 842 610	1 025 700	111 130				415 670	287 910	486 150	1 584 840	45 990	6 800 000
Cameroon	1 357 720	9 148 050	152 960	10 340	910		3 706 470	2 936 880	488 980	13 196 740	950	31 000 000
Central African Republic	819 700	2 878 090					13 930	209 760		857 520		4 779 000
Chad	1 648 610			13 310			2 637 880			900 200	0	5 200 000
Congo		449 020						143 030		1 803 190	4 760	2 400 000
Democratic Republic of the Congo	88 340	2 044 660	7 850				35 900	3 634 950	196 920	13 739 140	21 240	19 769 000
Equatorial Guinea	620	1 810	0		0		0	70 350	420	276 800		350 000
Gabon		149 820						162 520		2 786 820	840	3 100 000
Sao Tome and Principe							84 160	540		296 300		381 000
EASTERN AFRICA												
Burundi							30 890	939 370	3 151 720	178 020		4 300 000
Comoros								9 780		500 220		510 000
Eritrea	402 550		41 200	140			268 770		137 370	519 970		1 370 000
Ethiopia	2 724 750	64 810	229 240	10			10 623 110	1 219 450	22 634 340	1 504 290	0	39 000 000
Kenya	2 768 580		536 830				6 121 730	3 069 740	8 416 760	4 960 940	125 420	26 000 000
Madagascar	9 352 680	26 470	924 690				6 460 990	5 136 890	3 325 710	7 490 410	82 160	32 800 000
Malawi	668 630	204 050	126 970				10 929 640	487 940	307 580	2 454 470	20 720	15 200 000
Mozambique	11 257 940	158 260	32 100				9 015 830	110 060	12 170	8 082 560	1 080	28 670 000
Rwanda			0				299 610	231 250	1 378 330	90 810	0	2 000 000
Somalia	2 318 370		8 260	45 930			137 320			890 120	0	3 400 000
Uganda	1 658 980	1 059 240	109 560				3 563 660	21 196 000	3 379 070	2 033 490	0	33 000 000
United Republic of Tanzania	4 179 250	625 420	187 750				14 348 120	5 277 970	1 782 120	4 911 880	7 490	31 320 000

(Continued)

TABLE B.6 NUMBERS OF POULTRY BY COUNTRY AND LIVESTOCK PRODUCTION SYSTEM *(Continued)*

Region/Country	LGA	LGH	LGT	MIA	MIH	MIT	MRA	MRH	MRT	OTHER	URBAN	FAOSTAT 2005
EASTERN AFRICA *(Continued)*												
Zambia	15 403 090	220 420	190 570				9 475 610	60 510	9 600	4 223 380	416 820	30 000 000
Zimbabwe	5 837 230		66 130				13 206 310		1 462 880	2 323 220	201 230	23 097 000
SOUTHERN AFRICA												
Botswana	2 673 200						1 300 030			24 290	2 480	4 000 000
Lesotho	2 860		619 230						1 172 550		5 360	1 800 000
Namibia	2 434 180		1 850				877 350		390	121 400	64 830	3 500 000
South Africa	40 663 850	318 470	7 688 940	85 190			26 163 750	7 375 560	22 544 350	10 893 530	6 256 360	121 990 000
Swaziland	105 280						431 010	76 430	1 176 540	1 405 270	5 470	3 200 000

APPENDIX C

LIVESTOCK NUMBERS FOR CENTRAL, EASTERN, SOUTHERN AND SOUTH-EASTERN ASIA

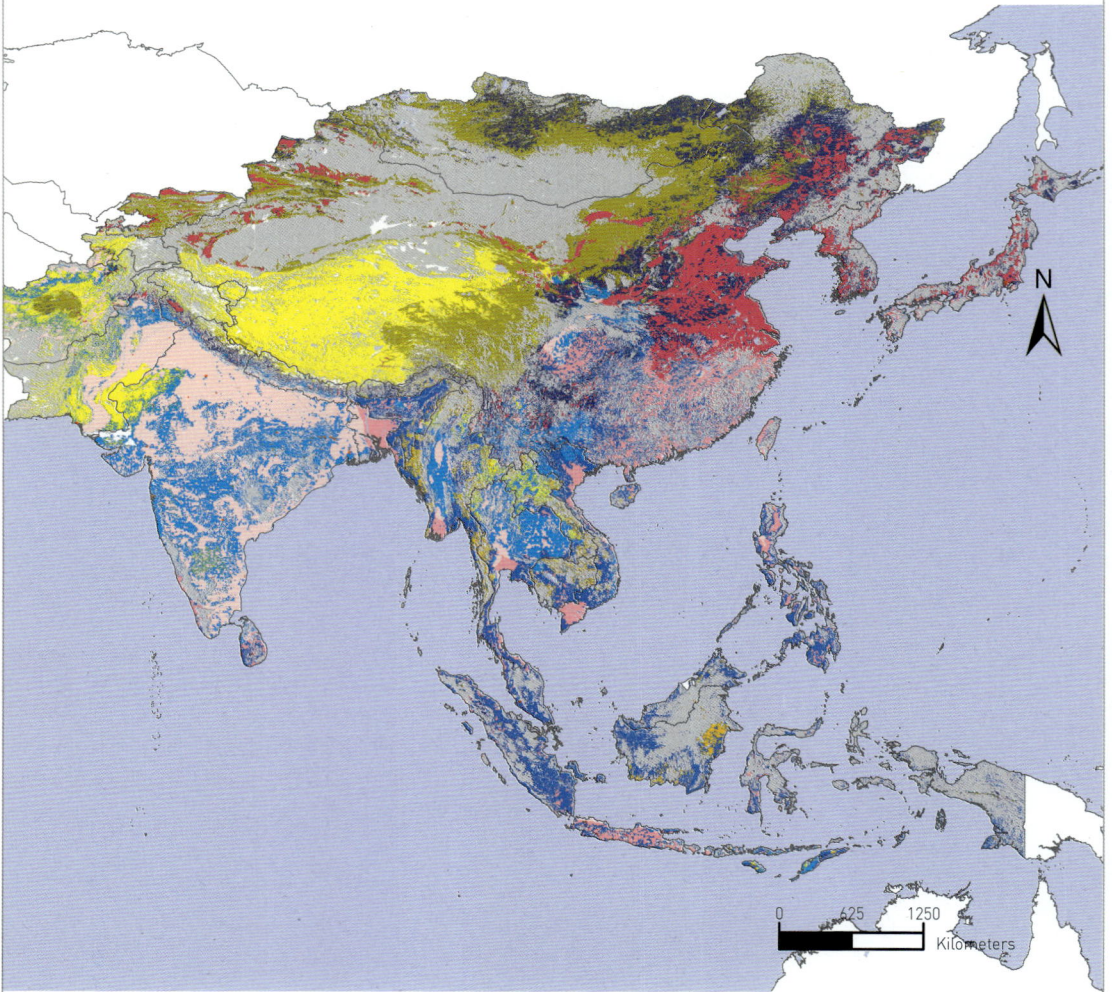

C1 LIVESTOCK PRODUCTION SYSTEMS IN CENTRAL, EASTERN, SOUTHERN AND SOUTH-EASTERN ASIA

Production systems

LGA	MIA	MRA	Other	International boundary
LGH	MIH	MRH	Urban	Water
LGT	MIT	MRT		

TABLE C.1 CATTLE NUMBERS BY COUNTRY AND LIVESTOCK PRODUCTION SYSTEM

Region/Country	LGA	LGH	LGT	MIA	MIH	MIT	MRA	MRH	MRT	OTHER	URBAN	FAOSTAT 2005
CENTRAL ASIA												
Kazakhstan	n.a.	n.a.	n.a.	n.a.	n.a.	n.a.	n.a.	n.a.	n.a.	n.a.	n.a.	5 181 000
Kyrgyzstan	17 450		509 100	2 850		261 410			18 060	226 020		1 034 890
Tajikistan	456 450		87 480	296 390		90 780	20 590		49 050	302 560		1 303 300
Turkmenistan	n.a.	n.a.	n.a.	n.a.	n.a.	n.a.	n.a.	n.a.	n.a.	n.a.	n.a.	2 024 500
Uzbekistan	n.a.	n.a.	n.a.	n.a.	n.a.	n.a.	n.a.	n.a.	n.a.	n.a.	n.a.	5 400 000
EASTERN ASIA												
China	936 150	13 920	7 163 980	1 612 350	6 915 970	35 670 250	4 250 840	6 899 860	25 929 710	25 832 170	4 300	115 229 500
Dem People's Rep of Korea			180	670		219 440	100		35 920	321 690		578 000
Japan	0		3 840	3 960	373 190	606 930	4 680	200 500	474 530	2 733 380	0	4 401 000
Mongolia			869 650			2 370			470 950	497 380	1 250	1 841 600
Republic of Korea			600	3 680		854 830	4 200	6 210	456 310	972 020	150	2 298 000
SOUTHERN ASIA												
Afghanistan	1 038 630		204 820	850 330		65 690	579 360		186 790	774 380		3 700 000
Bangladesh		17 830	0	8 280 860	12 463 360		1 168 910	2 445 870		119 100	4 070	24 500 000
Bhutan	4 220	840	10 260		8 910	13 050	0	78 960	97 050	158 710		372 000
India	2 090 190	64 920	58 320	88 130 790	2 639 310	91 570	54 123 840	7 952 820	1 055 490	28 218 570	574 180	185 000 000
Iran (Islamic Republic of)	n.a.	n.a.	n.a.	n.a.	n.a.	n.a.	n.a.	n.a.	n.a.	n.a.	n.a.	8 800 000
Nepal	16 860		8 770	2 311 070	4 470	106 010	1 321 290	30 250	550 840	2 642 070	2 820	6 994 450
Pakistan	2 758 260		54 000	14 452 540	6 910	407 040	2 902 420		440 300	3 168 820	9 710	24 200 000
Sri Lanka	10			175 820	245 260	3 760	174 270	298 390	11 760	306 930	1 800	1 218 000

(Continued)

TABLE C.1 CATTLE NUMBERS BY COUNTRY AND LIVESTOCK PRODUCTION SYSTEM *(Continued)*

Region/Country	LGA	LGH	LGT	MIA	MIH	MIT	MRA	MRH	MRT	OTHER	URBAN	FAOSTAT 2005
SOUTH-EASTERN ASIA												
Brunei Darussalam										1 300		1 300
Cambodia	7 210	158 770		15 140	372 780		123 030	1 870 320		552 750		3 100 000
Indonesia	19 960	56 920	3 280	121 070	2 467 570	241 270	467 980	3 609 620	167 730	4 334 820	9 780	11 500 000
Lao People's Democratic Republic	195 300	121 810	6 030	82 120	14 700		367 650	169 060	1 380	341 950		1 300 000
Malaysia	140	1 120	90		91 950		50	374 740	100	283 020	3 790	755 000
Myanmar	410 470	980 260	147 930	982 660	681 790	3 620	2 560 320	2 238 740	102 100	3 892 110	0	12 000 000
Philippines		2 220		107 760	574 240	1 540	46 430	1 526 060	18 090	313 070	1 590	2 591 000
Singapore								40		20	140	200
Thailand	63 470	35 960		1 401 710	471 270		2 232 350	702 390	260	591 690	900	5 500 000
Timor-Leste	16 590	25 400		50	0		13 290	81 520	8 000	26 150		171 000
Viet Nam	7 750	42 920	940	262 790	1 509 140	960	679 430	1 706 710	37 770	1 000 100	1 490	5 250 000

TABLE C.2 NUMBERS OF BUFFALOES BY COUNTRY AND LIVESTOCK PRODUCTION SYSTEM

Region/Country	LGA	LGH	LGT	MIA	MIH	MIT	MRA	MRH	MRT	OTHER	URBAN	FAOSTAT 2005
CENTRAL ASIA												
Kazakhstan	n.a.	n.a.	n.a.	n.a.	n.a.	n.a.	n.a.	n.a.	n.a.	n.a.	n.a.	9 000
EASTERN ASIA												
China	16 360	20 000	10 840	658 360	4 294 700	3 097 820	1 476 360	3 851 720	2 013 170	7 304 920	1 000	22 745 250
SOUTHERN ASIA												
Bangladesh		1 420		313 700	319 350		42 190	169 430	0	3 910	0	850 000
Bhutan	0	0	30		50	230	0	220	570	900		2 000
India	644 790	15 320	12 000	65 009 770	202 390	40 950	21 309 800	822 460	394 570	9 334 050	213 900	98 000 000
Iran (Islamic Republic of)	0			0	0	0	0			550 000	0	550 000
Nepal	11 030		3 260	1 217 760	7 790	89 170	801 360	27 350	309 310	1 614 420	0	4 081 450
Pakistan	854 400		2 980	22 103 680	2 310	120 800	1 669 520	0	284 780	1 254 130	7 400	26 300 000
Sri Lanka	100			54 470	43 950	1 180	65 730	64 130	2 010	84 000	430	316 000
SOUTH-EASTERN ASIA												
Brunei Darussalam										5 000		5 000
Cambodia	630	53 480		650	47 480		11 780	376 530	0	159 450		650 000
Indonesia	5 780	27 160	16 830	19 560	307 800	57 000	61 180	823 790	78 410	1 029 170	1 510	2 428 190
Lao People's Democratic Republic	176 800	143 020	6 780	32 620	10 860		258 830	161 300	1 880	337 910		1 130 000
Malaysia	0	1 090	10	0	9 080		0	65 390	40	53 670	720	130 000
Myanmar	102 050	227 100	39 790	257 850	74 060	390	603 500	461 300	28 480	905 480	0	2 700 000
Philippines		1 990		114 220	740 840	470	36 420	1 710 940	15 250	645 670	1 200	3 267 000
Singapore								10		0	10	20
Thailand	10 650	4 190		283 300	62 370		1 084 850	248 360	40	105 780	460	1 800 000
Timor-Leste	1 430	10 690		0	0		2 210	74 510	14 060	7 100		110 000
Viet Nam	11 170	8 570	16 560	32 160	874 830	300	359 080	1 223 640	98 660	339 940	0	2 950 000

TABLE C.3 NUMBERS OF GOATS BY COUNTRY AND LIVESTOCK PRODUCTION SYSTEM

Region/Country	LGA	LGH	LGT	MIA	MIH	MIT	MRA	MRH	MRT	OTHER	URBAN	FAOSTAT 2005
CENTRAL ASIA												
Kazakhstan	n.a.	n.a.	n.a.	n.a.	n.a.	n.a.	n.a.	n.a.	n.a.	n.a.	n.a.	1 995 300
Kyrgyzstan	16 590		339 600	2 450		189 530			10 300	249 920		808 390
Tajikistan	331 840		74 040	170 720		54 800	23 870		26 380	293 350		975 000
Turkmenistan	n.a.	n.a.	n.a.	n.a.	n.a.	n.a.	n.a.	n.a.	n.a.	n.a.	n.a.	822 000
Uzbekistan	n.a.	n.a.	n.a.	n.a.	n.a.	n.a.	n.a.	n.a.	n.a.	n.a.	n.a.	1 000 000
EASTERN ASIA												
China	7 495 720	13 180	30 838 980	2 350 440	4 296 670	70 133 830	4 299 810	5 134 380	38 083 540	33 100 200	12 200	195 758 950
Dem People's Rep of Korea			2 430	2 460		784 310	680		192 270	1 767 850		2 750 000
Japan	0		0	20	2 470	2 850	10	670	5 680	22 300	0	34 000
Mongolia			3 588 410			10 810			1 422 310	7 213 620	2 850	12 238 000
Republic of Korea			210	1 190		214 790	1 900	800	118 860	232 090	160	570 000
SOUTHERN ASIA												
Afghanistan	2 270 800	n.a.	273 250	1 044 100	n.a.	68 840	1 251 400	n.a.	326 760	2 064 870	n.a.	7 300 000
Bangladesh		35 940		15 884 440	16 547 300		1 707 380	2 448 410		276 460	70	36 900 000
Bhutan	400	180	690		1 300	1 500	0	10 060	5 910	9 960		30 000
India	3 634 990	18 210	37 340	59 319 630	1 021 670	56 650	36 183 390	3 595 240	672 320	15 079 140	381 420	120 000 000
Iran (Islamic Republic of)	n.a.	n.a.	n.a.	n.a.	n.a.	n.a.	n.a.	n.a.	n.a.	n.a.	n.a.	26 500 000
Nepal	46 600		16 320	2 107 560	7 410	114 220	1 608 890	33 230	530 720	2 686 580	2 000	7 153 530
Pakistan	11 374 670		50 580	28 191 290	17 180	439 780	7 022 020		710 780	8 870 240	23 460	56 700 000
Sri Lanka	680			52 660	126 380	1 550	54 980	90 730	4 000	93 640	380	425 000

(Continued)

TABLE C.3 NUMBERS OF GOATS BY COUNTRY AND LIVESTOCK PRODUCTION SYSTEM *(Continued)*

Region/Country	LGA	LGH	LGT	MIA	MIH	MIT	MRA	MRH	MRT	OTHER	URBAN	FAOSTAT 2005
SOUTH-EASTERN ASIA												
Brunei Darussalam										3 000		3 000
Indonesia	36 800	66 370	4 410	187 190	4 140 980	92 670	273 320	5 020 560	154 970	3 189 480	15 350	13 182 100
Lao People's Democratic Republic	30 400	17 430	1 010	1 090	400		40 160	12 860	160	39 490		143 000
Malaysia	150	830	60		28 130		50	102 220	60	91 400	2 100	225 000
Myanmar	48 900	143 970	31 800	241 300	52 370	120	419 190	272 510	23 870	565 970	0	1 800 000
Philippines		1 400		346 890	1 237 800	2 020	101 030	3 862 700	44 580	892 890	10 690	6 500 000
Singapore								120		60	420	600
Thailand	820	1 470		35 550	51 740		22 240	112 900	0	45 280	0	270 000
Timor-Leste	4 490	6 630		390	0		8 860	45 520	4 840	9 270		80 000
Viet Nam	4 490	6 010	230	57 960	253 580	230	203 100	440 370	21 830	212 140	60	1 200 000

119

TABLE C.4 NUMBERS OF SHEEP BY COUNTRY AND LIVESTOCK PRODUCTION SYSTEM

Region/Country	LGA	LGH	LGT	MIA	MIH	MIT	MRA	MRH	MRT	OTHER	URBAN	FAOSTAT 2005
CENTRAL ASIA												
Kazakhstan	n.a.	n.a.	n.a.	n.a.	n.a.	n.a.	n.a.	n.a.	n.a.	n.a.	n.a.	11 286 700
Kyrgyzstan	55 270	1 477 090		18 590	649 030				97 630	667 610		2 965 220
Tajikistan	593 310	170 230		234 310	103 730		47 020		68 790	564 610		1 782 000
Turkmenistan	n.a.	n.a.	n.a.	n.a.	n.a.	n.a.	n.a.	n.a.	n.a.	n.a.	n.a.	14 267 000
Uzbekistan	n.a.	n.a.	n.a.	n.a.	n.a.	n.a.	n.a.	n.a.	n.a.	n.a.	n.a.	9 500 000
EASTERN ASIA												
China	21 749 800	110	65 958 400	1 007 990	50 330	32 072 520	803 570	12 760	24 010 300	25 209 360	7 070	170 882 210
Dem People's Rep of Korea	0		220	160		31 130	10		21 750	118 730		172 000
Japan	0	700	0	0	120	3 250	0	20	1 570	6 040	0	11 000
Mongolia			4 774 840			17 470			2 011 570	4 878 000	4 520	11 686 400
Republic of Korea		580	0	0		190	0	10	330	460	110	1 100
SOUTHERN ASIA												
Afghanistan	2 937 700		484 360	1 864 630		62 620	1 598 450		229 570	1 622 670		8 800 000
Bangladesh		700		513 110	628 050		34 520	78 860		4 730	30	1 260 000
Bhutan	30	0	320		50	420		3 660	5 720	9 800		20 000
India	5 320 940		64 300	27 647 590	36 190	41 010	22 081 980	164 480	600 770	6 419 980	122 180	62 500 000
Iran (Islamic Republic of)	n.a.	n.a.	n.a.	n.a.	n.a.	n.a.	n.a.	n.a.	n.a.	n.a.	n.a.	54 000 000
Nepal	21 360		20 770	78 620	450	18 180	110 140	7 210	136 310	423 090	590	816 720
Pakistan	6 627 610		9 440	8 186 890	9 860	189 550	3 546 900	146 110	6 175 010	8 630	24 900 000
Sri Lanka	0			1 990	2 770	0	1 710	3 010	50	2 470	0	12 000

(Continued)

TABLE C.4 NUMBERS OF SHEEP BY COUNTRY AND LIVESTOCK PRODUCTION SYSTEM *(Continued)*

Region/Country	LGA	LGH	LGT	MIA	MIH	MIT	MRA	MRH	MRT	OTHER	URBAN	FAOSTAT 2005
SOUTH-EASTERN ASIA												
Brunei Darussalam										3 000		3 000
Indonesia	4 380	25 870	260	164 570	3 562 070	66 830	49 380	3 480 510	78 930	860 500	13 630	8 306 930
Malaysia	70	40	0		32 780		0	52 600	0	32 640	870	119 000
Myanmar	8 860	25 680	5 130	67 570	15 350	10	182 400	69 710	4 870	112 420	0	492 000
Philippines		10		420	12 490	30	90	9 320	170	7 460	10	30 000
Thailand	90	550		13 600	16 540		2 810	11 300	0	5 110	0	50 000
Timor-Leste	1 200	700		10	0		1 850	16 610	2 770	1 860		25 000

121

TABLE C.5 NUMBERS OF PIGS BY COUNTRY AND LIVESTOCK PRODUCTION SYSTEM

Region/Country	LGA	LGH	LGT	MIA	MIH	MIT	MRA	MRH	MRT	OTHER	URBAN	FAOSTAT 2005
CENTRAL ASIA												
Kazakhstan	n.a.	n.a.	n.a.	n.a.	n.a.	n.a.	n.a.	n.a.	n.a.	n.a.	n.a.	1 292 100
Kyrgyzstan	90	23 610		0	48 820				310	9 820		82 650
Tajikistan	130		20	20		170	30		0	330		700
Turkmenistan	n.a.	n.a.	n.a.	n.a.	n.a.	n.a.	n.a.	n.a.	n.a.	n.a.	n.a.	30 000
Uzbekistan	n.a.	n.a.	n.a.	n.a.	n.a.	n.a.	n.a.	n.a.	n.a.	n.a.	n.a.	90 000
EASTERN ASIA												
China	882 580	136 950	10 074 060	17 769 000	77 693 280	123 989 930	21 348 480	53 436 810	69 345 210	113 993 060	140 610	488 809 970
Dem People's Rep of Korea			820	4 800		1 389 260	800		207 620	1 596 700		3 200 000
Japan	0		1 180	7 840	1 064 130	1 518 080	35 780	596 770	820 850	5 505 370	0	9 550 000
Mongolia			2 240			0			1 360	2 400	0	6 000
Republic of Korea			1 730	21 850		3 076 480	16 620	15 130	1 662 920	3 775 150	120	8 570 000
SOUTHERN ASIA												
Afghanistan	0		0	0	n.a.	0	0	n.a.	0	0	n.a.	0
Bhutan	380	110	1 650		720	590	0	10 880	9 370	17 300		41 000
India	51 670	59 940	9 430	6 386 700	216 290	11 620	3 486 330	1 233 040	117 790	2 690 620	36 570	14 300 000
Iran (Islamic Republic of)	n.a.	n.a.	n.a.	n.a.	n.a.	n.a.	n.a.	n.a.	n.a.	n.a.	n.a.	0
Nepal	1 320		19 820	260 690	2 010	57 450	198 400	5 430	71 840	327 050	3 700	947 710
Sri Lanka	40			3 310	11 720	1 270	7 720	28 140	2 070	28 380	350	83 000

(Continued)

TABLE C.5 NUMBERS OF PIGS BY COUNTRY AND LIVESTOCK PRODUCTION SYSTEM *(Continued)*

Region/Country	LGA	LGH	LGT	MIA	MIH	MIT	MRA	MRH	MRT	OTHER	URBAN	FAOSTAT 2005
SOUTH-EASTERN ASIA												
Brunei Darussalam										1 800		1 800
Cambodia	6 040	135 510		4 640	304 960		91 670	1 412 310		544 870		2 500 000
Indonesia	79 730	105 960	2 100	35 020	661 380	28 880	611 110	1 923 450	133 600	2 668 710	17 430	6 267 370
Lao People's Democratic Republic	271 010	265 430	13 710	78 640	7 270		334 450	241 860	3 720	533 910		1 750 000
Malaysia	40	7 850	140		211 850		100	1 055 520	470	851 560	22 470	2 150 000
Myanmar	247 230	421 120	71 990	304 370	247 360	230	960 060	904 870	63 780	1 998 990	0	5 220 000
Philippines		4 350		181 510	2 580 200	6 960	168 540	6 592 400	42 620	2 545 990	16 430	12 139 000
Singapore	n.a.	n.a.	n.a.	n.a.	n.a.	n.a.	n.a.	n.a.	n.a.	n.a.	n.a.	250 000
Thailand	81 610	15 930		1 293 760	2 204 940		1 883 270	1 172 580	40	547 370	500	7 200 000
Timor-Leste	12 810	22 030		2 700	1 320		17 120	212 150	24 070	53 800		346 000
Viet Nam	29 900	156 990	5 620	472 820	13 152 750	1 760	1 654 040	7 915 600	232 770	3 372 900	4 850	27 000 000

TABLE C.6 NUMBERS OF POULTRY BY COUNTRY AND LIVESTOCK PRODUCTION SYSTEM

Region/Country	LGA	LGH	LGT	MIA	MIH	MIT	MRA	MRH	MRT	OTHER	URBAN	FAOSTAT 2005
CENTRAL ASIA												
Kazakhstan	n.a.	n.a.	n.a.	n.a.	n.a.	n.a.	n.a.	n.a.	n.a.	n.a.	n.a.	25 580 000
Kyrgyzstan	44 880	1 465 200		4 360	2 218 140				51 900	726 520		4 511 000
Tajikistan	620 850		181 320	532 660		347 550	32 910		101 260	479 650		2 296 200
Turkmenistan	n.a.	n.a.	n.a.	n.a.	n.a.	n.a.	n.a.	n.a.	n.a.	n.a.	n.a.	7 200 000
Uzbekistan	n.a.	n.a.	n.a.	n.a.	n.a.	n.a.	n.a.	n.a.	n.a.	n.a.	n.a.	18 350 000
EASTERN ASIA												
China	4 135 960	6 122 380	67 571 420	128 735 130	674 442 840	2 389 009 400	103 296 360	358 951 510	729 670 150	889 231 150	2 110 700	5 353 277 000
Dem People's Rep of Korea			3 870	29 090		12 125 450	7 610		1 705 240	12 628 740		26 500 000
Japan	0		23 260	420 730	27 014 070	45 807 170	633 930	17 410 270	24 292 920	167 400 650	0	283 003 000
Mongolia			9 590			10			8 440	11 960	0	30 000
Republic of Korea			19 370	592 580		47 755 940	526 660	547 540	23 190 300	46 391 470	140	119 024 000
SOUTHERN ASIA												
Afghanistan	2 075 750		371 750	2 295 730		113 770	1 255 480		321 680	1 965 840		8 400 000
Bangladesh		84 220		40 630 770	81 959 530		8 034 210	21 832 270		1 071 290	87 710	153 700 000
Bhutan	1 790	870	6 260		8 120	7 870	0	73 880	46 620	84 590		230 000
India	2 764 050	1 055 670	213 950	205 738 790	21 096 140	232 800	113 051 590	31 504 750	1 864 650	83 079 220	2 398 390	463 000 000
Iran (Islamic Republic of)	n.a.	n.a.	n.a.	n.a.	n.a.	n.a.	n.a.	n.a.	n.a.	n.a.	n.a.	284 600 000
Nepal	85 030		57 560	8 900 280	15 150	412 700	4 377 500	86 690	1 751 910	7 226 130	277 050	23 190 000
Pakistan	18 845 620		908 330	90 023 540	47 120	1 926 360	27 696 130		3 963 220	25 953 650	136 030	169 500 000
Sri Lanka	5 020			674 510	1 727 930	195 770	1 127 130	3 830 050	429 100	3 535 490	93 000	11 618 000

(Continued)

TABLE C.6 NUMBERS OF POULTRY BY COUNTRY AND LIVESTOCK PRODUCTION SYSTEM *(Continued)*

Region/Country	LGA	LGH	LGT	MIA	MIH	MIT	MRA	MRH	MRT	OTHER	URBAN	FAOSTAT 2005
SOUTH-EASTERN ASIA												
Brunei Darussalam										13 100 000		13 100 000
Cambodia	69 720	1 116 140		58 830	2 623 640		801 450	13 078 170		4 252 050		22 000 000
Indonesia	745 550	7 262 990	327 730	35 633 650	395 404 160	11 707 610	9 843 680	494 596 180	23 236 130	301 978 840	2 964 480	1 283 701 000
Lao People's Democratic Republic	3 685 210	2 204 010	95 210	2 588 680	321 110		6 162 190	3 441 100	37 920	5 764 570		24 300 000
Malaysia	44 500	803 920	52 170		16 229 030		14 480	109 649 580	31 910	71 816 220	2 358 190	201 000 000
Myanmar	4 588 520	8 866 580	1 250 450	4 782 270	3 793 600	1 280	16 707 520	17 678 150	762 070	38 231 560	0	96 662 000
Philippines		72 960		4 354 100	41 874 560	116 390	1 708 600	75 428 810	498 810	23 046 210	359 560	147 460 000
Singapore	n.a.	n.a.	n.a.	n.a.	n.a.	n.a.	n.a.	n.a.	n.a.	n.a.	n.a.	2 600 000
Thailand	1 056 720	1 071 990		53 391 850	53 161 070		82 036 300	66 924 300	0	19 609 230	18 540	277 270 000
Timor-Leste	188 950	114 140		11 700	13 760		269 960	1 117 930	62 580	420 980		2 200 000
Viet Nam	112 730	955 490	7 740	3 731 590	123 674 630	24 840	17 723 580	73 059 640	1 298 870	24 335 740	75 150	245 000 000

TABLE D.1

Continent/Region/Country	Cattle	Buffaloes	Sheep	Goats	Pigs	Poultry
AMERICAS						
Northern America						
Canada	15 083 000	0	1 000 000	30 000	14 675 000	167 050 000
United States of America	95 848 000	0	6 135 000	2 522 500	60 644 500	2 044 900 000
ASIA						
Western Asia						
Armenia	573 300	402	558 300	45 000	89 100	3 699 400
Azerbaijan	2 007 206	308 551	6 887 444	601 387	22 932	18 253 000
Cyprus	57 000	0	295 000	460 000	498 000	3 870 000
Georgia	1 250 700	35 000	689 200	115 700	483 900	9 836 200
Iraq	1 500 000	120 000	6 200 000	1 650 000	0	33 000 000
Israel	357 000	0	435 000	65 000	200 000	35 795 000
Jordan	69 100	100	1 671 535	444 450	0	25 013 000
Kuwait	28 000	0	900 000	150 000	0	32 500 000
Lebanon	90 000	0	340 000	430 000	15 000	35 000 000
Oman	335 000	0	375 000	1 070 000	0	4 200 000
Saudi Arabia	350 000	0	7 000 000	2 200 000	0	141 000 000
Syrian Arab Republic	940 000	2 800	15 310 000	1 018 000	0	30 350 000
Turkey	10 069 346	103 900	25 201 156	6 609 037	4 399	302 978 000
United Arab Emirates	115 000	0	580 000	1 520 000	0	15 000 000
West Bank	34 000	0	800 000	400 000	0	7 000 000
Yemen	1 400 000	0	6 600 000	7 300 000	0	37 000 000
EUROPE						
Eastern Europe						
Belarus	3 962 600	0	59 000	66 000	3 406 800	25 100 000
Bulgaria	671 579	7 973	1 692 507	718 117	931 402	19 140 000
Czech Republic	1 397 308	0	140 197	12 623	2 876 834	13 690 000
Hungary	723 000	0	1 397 000	78 000	4 059 000	41 330 000
Moldova, Republic of	331 000	0	823 000	119 000	397 000	17 522 000
Poland	5 483 290	0	315 963	0	18 112 380	98 100 000
Romania	2 812 000	0	7 430 000	662 000	6 589 000	98 455 000
Russian Federation	22 987 700	17 288	15 493 719	2 277 366	13 412 770	334 708 000
Slovakia	580 000	0	316 000	40 000	1 300 000	13 230 000
Ukraine	6 952 700	0	875 200	894 300	6 466 100	152 800 000
Northern Europe						
Denmark	1 544 296	0	160 745	0	13 466 283	17 350 000
Estonia	249 800	0	38 800	2 900	340 100	2 183 000

(Continued)

TABLE D.1 *(Continued)*

Continent/Region/Country	Cattle	Buffaloes	Sheep	Goats	Pigs	Poultry
EUROPE						
Northern Europe						
Faroe Islands	2 000	0	68 100	0	0	0
Finland	950 000	0	115 000	7 300	1 365 000	6 000 000
Iceland	64 000	0	454 000	410	35 000	190 000
Ireland	7 000 000	0	4 556 700	7 700	1 757 600	14 595 000
Latvia	371 100	0	38 600	14 700	435 700	4 050 000
Lithuania	791 966	0	22 149	26 904	1 073 348	8 418 230
Norway	920 300	0	2 417 000	64 500	515 400	3 300 000
Sweden	1 619 000	0	479 400	0	1 823 474	6 800 000
U.K. of Great Britain and Northern Ireland	10 378 023	0	35 253 048	0	4 851 000	159 845 000
Southern Europe						
Albania	700 000	120	1 800 000	950 000	140 000	6 475 000
Bosnia and Herzegovina	440 000	13 000	900 000	0	600 000	9 700 000
Croatia	471 025	0	796 480	120 000	1 205 000	11 541 000
Greece	600 000	788	9 000 000	5 400 000	1 000 000	28 193 000
Italy	6 314 000	237 000	8 200 000	985 000	9 272 000	126 000 000
Malta	17 900	0	14 900	5 400	73 000	1 010 000
Portugal	1 443 000	0	5 500 000	547 000	2 348 000	42 000 000
Serbia and Montenegro	1 254 000	28 700	1 828 000	192 000	3 189 000	17 521 000
Slovenia	451 136	0	94 000	22 000	533 998	5 431 000
Spain	6 700 000	0	22 500 000	2 750 000	25 250 000	130 902 000
The former Yugoslav Republic of Macedonia	248 185	725	1 244 000	0	155 753	2 617 000
Western Europe						
Austria	2 051 000	0	327 200	55 500	3 125 400	12 462 000
Belgium	2 694 662	0	155 333	26 455	6 332 433	34 403 000
France	19 383 000	0	9 185 475	1 212 590	15 020 198	243 913 000
Germany	13 034 500	0	2 642 400	170 000	26 858 000	123 087 000
Liechtenstein	6 000	0	2 900	280	3 000	0
Luxembourg	184 172	0	7 500	2 000	84 547	72 828 000
Netherlands	3 862 000	0	1 236 000	282 000	11 153 000	87 900 000
Switzerland	1 540 000	0	443 000	74 000	1 594 000	8 193 000

(Continued)

TABLE D.1 *(Continued)*

Continent/Region/Country	Cattle	Buffaloes	Sheep	Goats	Pigs	Poultry
OCEANIA						
Australia and New Zealand						
Australia	27 730 000	0	102 700 000	400 000	2 490 000	87 220 000
New Zealand	9 609 000	0	39 928 000	155 000	341 000	20 325 000
Melanesia						
Fiji	310 000	0	5 000	260 000	140 000	4 390 000
New Caledonia	111 000	0	2 300	8 100	25 500	600 000
Papua New Guinea	91 500	0	7 500	2 700	1 750 000	4 026 000
Solomon Islands	13 500	0	0	0	53 000	230 000
Vanuatu	152 000	0	0	12 000	62 000	340 000
Micronesia						
Guam	130	0	0	680	5 100	205 000
Kiribati	0	0	0	0	12 400	460 000
Micronesia (Federated States of)	13 900	0	0	4 000	32 000	0
Nauru	0	0	0	0	2 800	5 000
Polynesia						
American Samoa	103	0	0	0	10 500	38 000
Cook Islands	120	0	0	1 000	32 000	15 000
French Polynesia	12 000	0	440	16 500	27 000	232 000
Niue	112	0	0	0	2 000	15 000
Samoa	29 000	0	0	0	201 000	450 000
Tokelau	0	0	0	0	1 000	5 000
Tonga	11 250	0	0	12 500	81 000	300 000
Tuvalu	0	0	0	0	13 500	60 000
Wallis and Futuna	60	0	0	7 000	25 000	63 000